SHARK ATTACKS

Myths, Misunderstandings and Human Fear

I0878569

BLAKE CHAPMAN

So that the oceans may always be wild, wonderful and full of mystery for Clint, Hannah and Boston.

And for C.M.B. and M.A.H.

SHARK ATTACKS

Myths, Misunderstandings and Human Fear

BLAKE CHAPMAN

PUBLISHING

National Library of Australia Cataloguing-in-Publication entry

    Chapman, Blake, author.

    Shark attacks : myths, misunderstandings and human fear / Blake Chapman.

    9781486307357 (paperback)
    9781486307364 (ePDF)
    9781486307371 (epub)

    Includes index.

    Sharks.
    Shark attacks.
    Dangerous marine animals.

Published by

CSIRO Publishing
Locked Bag 10
Clayton South VIC 3169
Australia

Telephone: +61 3 9545 8400
Email: publishing.sales@csiro.au
Website: www.publish.csiro.au

Front cover: Reef shark. Source: cdascher/istockphoto.

Set in 12/15 Adobe Garamond Pro and Myriad Pro
Edited by Adrienne de Kretser, Righting Writing
Cover design by Andrew Weatherill
Typeset by Desktop Concepts Pty Ltd, Melbourne
Printed in the USA by Integrated Books International

Foreword

Sharks – it seems like people either love them or hate them. How is it that one animal can invoke such vastly contrasting emotions?

As a shark attack survivor and now an environmentalist, I've been on both sides of the fence, or fin, so to speak. At one point in my life, I was of the belief that we would be better off if sharks were all killed, but now I know better. All it took was to be nearly killed by one. That was a seriously bad day at work. But what it did was force me into a realm of understanding. An understanding of the choices I made to put myself into that situation and an understanding of the risks of other choices I make every day, like riding a high powered motorcycle – something with far higher odds of getting me injured or killed than the risk of being attacked by a shark. It was my lifestyle and my choice of career as a navy diver that led me to the point of fighting for my life against a bull shark in Sydney Harbour. So how can I blame the shark?

We, as humans, have a duty of care. Not just for our own wellbeing but for our home: Earth. If we think about the stability of our planet's environment as being like a game of Jenga, it becomes clearer how important each piece is. Remove a piece here or there and the structure is still pretty solid. But if we continue to pollute our seas, pump toxins into the earth and poison our air then where is the tipping point? Our sharks are a vital piece in this game and we need to understand that the safety of humans in the water cannot come entirely at the expense of sharks.

Shark Attacks emphasises the fact that human–shark interactions are complex. While not diminishing the devastating realities of shark attacks, Blake Chapman provides a balanced account of the facts and the fiction around shark attacks – not just what's in the media or common 'knowledge'.

Knowledge is and always will be power. Knowledge is the reason that I can now dive with great white sharks without the safety of a cage. It is the reason I can share the underwater realm 100 feet down in harmony with massive schools of sharks and come out unscathed and even feed bull sharks by hand – not with a hand. Which is a pleasant change.

This book is a wonderful example of the power that comes through knowledge. My experience is a testament to the fact that when we better understand our emotional responses to sharks and take personal responsibility for our choices, we not only help to lower the odds of shark attack even further but also gain a richer appreciation for the world we live in.

Paul de Gelder
Navy diver, shark attack survivor, author and motivational speaker

Contents

Acknowledgements

First and foremost, the biggest of thanks to Clint Chapman. This is not only for the many discussions on content material and for reading draft chapters of the book, but mostly for supporting me in so many ways during writing – the book would not have been possible without you. Thanks also go to Dr Cora Lau for reviewing a draft and providing very useful comments and suggestions, and to those who very generously volunteered their time to help me to more fully understand various aspects of this material, especially Professor Ottmar Lipp and Dr Richard Cash. Thanks to Duncan Leadbitter and the Australian Aerial Patrol for allowing me to tag along on a few shark-spotting flights. I also greatly appreciate the support of so many of my family and friends, who have had to listen to me go on and on about sharks and shark attacks for so long. Not only did you listen, you also provided excellent suggestions and feedback and helped to keep me motivated. We can change the subject now. Many thanks to the CSIRO Publishing team for your assistance, guidance and support along the way, especially Briana Melideo, Lauren Webb and Tracey Kudis. Last, but certainly not least, I would like to offer a huge degree of gratitude to all the people who took the time to speak to me about their experiences relating to shark attacks, especially those who contributed personal accounts. This is something that scientific research cannot provide, and the new perspectives I gained from speaking to you have certainly opened my eyes to the reality of the subject and transformed my views on many aspects of the topic of shark attacks. Statistics and science can tell us a lot, but you continuously reminded me that for as much as science can inform, it cannot provide the human aspect of the story. I know that these discussions would have been difficult for many of you – I acknowledge and thank you for your bravery and for your resolve to use your experiences to help educate others. You are truly inspirational people.

Prologue

Shark attacks are a highly emotive topic, on which people tend to have very intense and definite opinions. Therefore, it is acknowledged straight away that there may be points or sections within this book that you do not agree with. However, the book has been very carefully written to utilise current, and currently accepted, research. Shark attacks are big news events and attract media attention of all degrees and modalities. As a result, there is a wealth of coverage on the topic – but unfortunately, much of this information is partly or largely erroneous. Yet, as a result of not knowing otherwise, and often having information presented to us from only a singular angle, we tend to believe that information. Therefore, I encourage you to read this book with an open mind. Whether you are a novice or expert on the topic, some information within may be considered common knowledge, whereas other information may be considered extreme or incorrect. However, all information presented here is scientifically backed. That being said, I have also endeavoured to lighten the heavy load of scientific fact with stories of my own experience with sharks, as well as the stories of others. All of these stories are true, and they demonstrate both the lighter and more serious sides of interacting with sharks. The accounts that have been included are intended to provide a better understanding of shark attacks for those of us who have not been directly affected by such events. I would like to emphasise that the views and opinions in these accounts are those of the individuals who have provided them. They are designed to not only inform about the intimate details of shark attacks, but also to reflect and provide the scope of the wide-ranging views of the general public.

Sharks are fascinating creatures, and the terror they invoke in our psyche is only a minor characteristic of what makes them such intriguing animals. We humans owe a great deal of thanks to sharks, which have significantly advanced our lives in a variety of fields, including engineering, tourism, medicine, pharmacology and ecology.

Not to mention the immense, yet rudimentarily understood, role they play in the functioning of the planet we inhabit. Although humans have rapidly developed the ability for higher cognitive function, we are but a blip in evolutionary history compared to sharks. It is up to us to learn how to better appreciate these animals for what they are, what they do for the aquatic environment, and how they function. And it is up to us to learn how to coexist with them, not the other way around. Although the risk of shark attack on humans is incredibly low, it is a significant concern to many people. And the fact remains that human lives are lost to sharks every year. Thus, more effective, well-developed mitigation methods are warranted and incredibly important. So is education. It is not unreasonable to believe that we will soon come up with methods to reduce shark–human interactions. Indeed, this is already underway and we are starting to have some technological and conceptual wins. However, basic and fundamental education is critical. It is hoped that this book may serve as a step in the right direction.

While I don't claim to have all the answers, I feel that I am at least qualified to present these topics for your consideration. I developed a keen interest in sharks when I was in my teens. From that time on, I have made it a point to learn more about sharks to help to address some of the unknowns and to gain a better understanding of these animals. This mission took me from the US to Australia, where I completed my Honours and PhD degrees in shark neuroscience, ecology, breeding and reproduction. My postgraduate research involved a wealth of laboratory and field research on sharks and rays. Upon the completion of my PhD, I worked at a large public aquarium on Queensland's Sunshine Coast. I was initially a diver and aquarist, then took on the role of Animal Health Manager. I worked hands-on with the entire animal collection, including the sharks, on a daily basis. I have since continued to research shark physiology and the science of shark attacks. My ultimate goal is to encourage both the conservation of sharks and the ability for humans to enjoy recreational water activities without the fear of being attacked by a shark.

Introduction

When I first became interested in sharks, I remember hearing somewhere (most likely on an episode of the Discovery Channel's infamous *Shark Week* series) that humans are generally safe from shark attacks because the human body does not contain enough fat and nutrition to sustain the effort expended by a shark during an attack. The documentary also informed viewers that sharks generally make an inquisitory bite before they attack. Through this process, they can instantaneously gauge the fat content of what they have bitten into and decide if it is worth the risk and energy expenditure for further effort. I was also informed that white sharks (*Carcharodon carcharias*), in particular, are actually shy, cautious predators, easily scared off by dangerous prey. As a largely naïve teenager, I was set. I had all the facts I needed. Why were people so scared of sharks? They weren't dangerous, they were simply misunderstood! And, armed with my 'degree' from Discovery Channel University, it became my immediate job to spread the word. And I did just that. I had to give a speech for my freshman year English class, so I presented on the topic of sharks and how they were simply misunderstood. Sure, I did some research into the subject, but really, I knew it all already. And after all, I was being assessed on my oral presentation skills, not the scientific accuracy of the information I was presenting. My mission to remedy the misunderstanding surrounding sharks did not stop there. Fortunately, my education on the topic did not end there either.

While there is a great deal of myth and misunderstanding around the topic of shark attack, the very succinct facts of the matter are that sharks do occasionally bite people, that sometimes, albeit rarely, these bites cause horrific and debilitating injuries, mental trauma and/or fatality, and that the global prevalence of shark bites is increasing. And while the reality of the situation can be so briefly summarised, to truly understand the occurrence of shark attacks, a much deeper knowledge base and an open mind are necessary. Thus, in this book, I provide a summary of the current research-based knowledge of sharks and shark attacks, the human fear of sharks, and the successes and failures of past and current mitigation measures and legislation. My goal is to comprehensively analyse the 'phenomenon' of shark attacks from a variety of angles, while dispelling myths on the topic (such as my previous 'knowledge' that white sharks are not dangerous to humans), so that we can start to form educated opinions on the subject. I don't mind what opinions people come away with at the end of this book, as long as they are based on fact not fiction, on reality not emotion, and on science not gossip.

1

Introduction to sharks

The word 'shark' has the potential to invoke fear in a way that few other things can. Many people think that sharks are large-bodied, man-eating predators that patrol the beaches waiting for their next meal. However, in addition to being blatantly incorrect, this definition grossly oversimplifies the diversity and complexity of animals that we identify as sharks. In fact, the most recent, comprehensive assessment tallied the currently described number of sharks at 509 species, covering nine orders, 34 families and 105 genera (Weigmann 2016). Sharks may be small, very large or anywhere in between, they range from planktivores to apex predators and they occupy a suite of environments including freshwater rivers, shallow tropical coastal waters, temperate deep water and arctic oceans.

Sharks and their relatives (rays, skates and chimaeras) comprise the Class Chondrichthyes, or cartilaginous fish, as their skeletons are made of cartilage not bone. Members of this Class have occupied the planet for longer than most other groups of animals in existence today. They are the second-oldest extant class of vertebrates, pre-dated only by the jawless fish (living examples of these are hagfish and lampreys), and the oldest extant clade of jawed vertebrates. Fossil records date the earliest Chondrichthyans back 400–450 million years. They quickly radiated out to populate the upper tiers of aquatic food webs and, based on evolutionary survivability, Chondrichthyans could be considered the most successful of all fish (Grogan and Lund 2004; Dulvy *et al.* 2014). Relatives of modern sharks were occupying the oceans for hundreds of million years before pterosaurs patrolled the skies and *Tyrannosaurus rex* scavenged the plains. And sharks survived when those species did not. Sharks most certainly pre-date all species of mammals, with an approximate 385 million-year jump on the evolution of higher primates (give or take 25 million years). While the evolutionarily oldest Chondrichthyan species are no longer in existence, their close relatives are. The extinct megatooth shark (*Carcharocles/Carcharodon megalodon*) is thought to be the largest shark (and predatory fish) ever recorded, with a body length of up to 18 m, and it potentially had a substantial impact on the evolutionary history of marine mammals (Lindberg and Pyenson 2006; Ferretti *et al.* 2010; Pimiento and Balk 2015). Today, Chondrichthyans represent one of the most species-rich lineages of predators on Earth and they continue to play important functional roles in coastal and oceanic ecosystem structure and function.

Love them or hate them, sharks are undeniably fascinating. Sharks include some of the latest to mature and slowest to reproduce vertebrates, have the longest gestation periods and are among the most dedicated in terms of gestational maternal investment (Cortés 2000). Whale sharks (*Rhincodon typus*) are the largest fish in the sea, capable of attaining a total length of more than 12 m and a weight of over 21 tonnes. The Greenland shark (*Somniosus microcephalus*), which can attain lengths of up to 5 m, is the longest-lived vertebrate known, with recent studies estimating a lifespan of about 390 years, give or take 120 years (Nielsen *et al.* 2016). The breadth and intricacies of shark reproduction are fascinating; some species give birth to live young, whereas others lay eggs. And of course there are the teeth. Sharks serially produce and shed their infamous teeth, with some species cycling through as many as 6000 teeth per year and many tens of thousands over their lifetime. Ultimately, the sheer diversity among shark species – their habitat range, variation in size and shape, biology and physiology, predation and social behaviours – makes this group of animals incredibly complex and captivating.

Our evolving interest in sharks

Although sharks represent one of the oldest vertebrate lineages in existence, we still have a lot to learn about them. Sharks were revered as gods (both good and devil-gods) in various ancient cultures around the world. However, the incredible interest that many people have in sharks today is a relatively recent development. Sharks didn't stand in the way of cultural expansion, exploration or industrialisation. Relative to terrestrial predators, they were not commonly encountered and (with some minor exceptions) did not comprise a commercially valuable food or consumable resource, as did whales, marine turtles, sea otters and certain bony fish species (Martin 2016). Although shark attacks occurred, and were of occasional regional interest in a few places during the first half of the 20th century, these were generally very isolated events and were rarely afforded any thought beyond the admission that an accident had taken place. The first major turning point in modern concern about sharks was the catastrophic sinking of the US naval warship, the USS *Indianapolis*, by a Japanese submarine in 1945. The ship had just delivered the crucial components of the first atomic bomb that would level Hiroshima days later. On 30 July, it was hit by two torpedoes and sank in 12 minutes. The 900 men who survived the sinking were left floating in the Pacific Ocean for four days, subject to exposure, thirst and sharks. The main species of sharks implicated was the oceanic whitetip (*Carcharhinus longimanus*). It is written that sharks consumed dozens to hundreds of bodies, but we don't know what proportion of those people were already dead before their consumption and how many were actually killed by the sharks. Either way, the fear felt by the sailors would have been incontrovertible. However, much of the detail surrounding this occurrence was classified at the time, and thus not as wide-reaching as it could have been. The resulting interest in sharks from the US Navy, however, prompted a scientific symposium on sharks in 1958, where 34 of the world's pre-eminent shark scientists came together to address the 'shark hazard problem' (Martin 2016). The report from that meeting stressed the overwhelming lack of knowledge in the field and the basic taxonomic and technical problems, such as the lack of standardised common names that would be needed for collaborative data collection (Aronson and Gilbert 1958).

The 1970s marked the start of significant and rapid change in shark interest and accompanying changes in human perceptions of sharks. This has taken us from ignorance, to fear and hatred and now to celebration and concern (although still with remnants of misunderstanding, fear and hatred). In a few short decades, sharks have gone from an 'out of sight, out of mind' inconsequential ocean resident to predatory monsters to an iconic symbol of troubled aquatic ecosystems and one of the most complex challenges to conservation, environmental management and regional human recreational safety.

The 1975 release of the movie *Jaws* was probably the single most influential event in shark–human history. The film managed to play on our innate fear of the unknown and uncontrollable, and simplified an entire, evolutionarily strong, biologically diverse and overwhelmingly complex group of animals into a single terrifying body part. It led to the widespread and non-specific destruction of huge numbers of sharks, whether they were potentially dangerous or not. Simultaneously, though, the movie also stimulated interest, curiosity and a whole new realm of research, as it brought sharks into the minds and living-rooms of the '*Jaws* Generation' in a big way.

Our more recent interest in sharks has shifted towards gaining a better understanding of these animals overall, their role in aquatic ecosystems and the conservation of threatened populations. The ecological, biological and physiological diversity of sharks leaves them susceptible to threats that are similarly wide-ranging and extinction risks to sharks are substantially higher than for most other vertebrate groups. Only one-third of shark species are considered to be safe from human exploitation and action (Dulvy *et al.* 2014). While the full degree to which sharks have been affected by human activity is not clear, and is often highly contentious among researchers, the most pressing current threats include the fin trade, habitat destruction and pollution, overfishing and climate change. Overfishing and habitat destruction have contributed to some of the most profound impacts, with 96.1% of sharks threatened by fishing. This includes direct commercial efforts (31.7%), bycatch (57.9%), recreational (0.7%) and artisanal fishing (5.8%). While not as significant, a further 2.9% are considered to be affected by habitat destruction and 0.4% by pollution (http://www.redlist.org).

Population trends

Studies from the 1970s to 2000s showed a generally grim outlook on shark population trends. During this time, fisheries logbooks and long-term shark mitigation beach mesh netting programs were most commonly used to study catch rates, which were then used to assess population trends. Results from netting programs in South Africa and Australia showed declines in bull (*Carcharhinus leucas*), blacktip (*Carcharhinus limbatus*), scalloped hammerhead (*Sphyrna lewini*), grey nurse (*Carcharhias taurus*) and white shark (*Carcharodon carcharias*) populations of 27% to more than 99% (Holden 1977; Paterson 1990). Early data from pelagic fisheries suggested that one shark was caught as bycatch for every two tuna, and two to three sharks for every swordfish (Brodie and Beck 1983; Baum and Myers 2004). Reductions in catch rates of 49% to more than 99% were recorded over less than 15 years for 20 species of coastal and pelagic sharks in the north-west Atlantic. Fourteen species of sharks disappeared completely from Brazilian fisheries between 1977 and 1994 (Amorim *et al.* 1998; Baum *et al.* 2003).

While population assessments based on net and fisheries catch data are subject to various biases and cannot be used to definitively determine population sizes, such significant figures clearly suggest a declining trend in many species. However, an interesting concomitant increase of shortfin mako (*Isurus oxyrinchus*) and blue sharks (*Prionace glauca*) was observed in Brazilian waters (Amorim *et al.* 1998), as well as blue shark biomass in the north Pacific, which was estimated to have increased by 20% relative to the 1970 numbers (Sibert *et al.* 2006). As large coastal species declined, catch rates of more wide-ranging species, such as tiger (*Galeocerdo cuvier*) and hammerhead sharks (*Sphyrna* spp.) increased, at least temporarily, as reported from the shark nets in New South Wales and Queensland (Australia) and South Africa (Paterson 1990; Reid and Krogh 1992; Dudley and Simpfendorfer 2006).

The longest dedicated population survey of sharks in the US has been running off the North Carolina coast since 1972. This survey has shown declines of up to 99% in a range of species, including bull, dusky (*Carcharhinus obscurus*) and tiger sharks (Myers *et al.* 2007). It also found that the average length of sharks decreased by 17–47%, functionally removing a large percentage of mature individuals from the populations.

Not only have indirect fishing pressures (through bycatch) increased, but directed shark fishing efforts have also increased, as more typically targeted bony fish species have become less accessible due to reduced stock or increased management restrictions. The value of shark products, including fins, liver, meat and gills, has also increased, due to greater demand, tougher legislation and decreased stock (Clarke *et al.* 2006; Lack and Sant 2009). Sharks are caught in at least 126 countries around the world, although the top 20 countries are responsible for 80% of the global annual catch (Lack and Sant 2011; Davidson *et al.* 2016). In total, the regulated legal global annual trade of shark products was estimated to be around US$1 billion in 2010 (FAO FAD 2010). Instead of reducing directed shark catch rates, initiatives to slow the shark fin trade, such as requiring the entire animal to be landed (i.e. brought on board the ship), encouraged greater utilisation of the whole animal, turning sharks into an even greater commercial resource. Estimates of global reported and unreported shark catches, including landings, discards and finning, were 1.44 million tonnes in 2000 and 1.41 million tonnes in 2010 (Worm *et al.* 2013). This equated to mortality estimates of 97–100 million sharks per year. Just to reiterate, that figure is 97–100 *million* sharks killed. *Per year.* Those numbers astound me.

The increase in targeted commercial shark fishing and the wider and more profitable use of shark products has resulted in the targeting of previously unexploited and untouched populations at greater distances to consumer markets (Dent and Clarke 2015). Many current levels of shark catches are unsustainable, and will need to change if we expect to keep these animals in existence. Fishing exploitation rates currently sit at 6.4–7.9% of shark populations annually, while the average rebound rate for sharks (as a whole, derived from assessment of the life history characteristics of 62 species) is 4.9% per year (Worm *et al.* 2013). Thus, most species affected by targeted and non-targeted fisheries should be expected to show continual declines. The most recent assessment by the International Union for Conservation of Nature (IUCN) suggests that of the 465 shark species investigated, 74 (15.9%) were considered threatened, including 11 (2.4%) categorised as Critically Endangered, 15 (3.2%) as Endangered and 48 (10.3%) as Vulnerable. A further 67 (14.4%) were assessed as Near Threatened and 209 (44.9%) as Data Deficient. Only 115 (24.7%) were considered of Least Concern (Dulvy *et al.* 2014).

Counting sharks

Despite their seemingly ever-present status along highly populated beaches, the reality is that sharks are extremely difficult to find, let alone count. The two most common methods used for population assessments are tagging and tracking and photo identification (typically based on fin contours or spot patterns, which are both as unique as our fingerprints), in conjunction with statistical modelling. However, one of the largest challenges facing population studies is the fact that it is simply beyond our current understanding to know exactly what percentage of a population may be present in a certain location (and thus subject to sampling) at a particular time. Sharks can be highly mobile and widely dispersed, unpredictable in their movements and residency patterns, and cryptic. Even population studies conducted over multiple years cannot guarantee that entire populations have been sampled. As a result, population assessments can be unreliable and are often highly contended among researchers. While some studies have found increasing trends in certain shark populations (e.g. tiger sharks off KwaZulu-Natal, South Africa, and white sharks off south-east Australia: Dudley and Simpfendorfer 2006; Reid *et al.* 2011), most studies indicate that shark populations are decreasing or stagnant. Recent studies have shown a decreasing trend in maximum body size of white, tiger and bull sharks in the last century (Kock and Johnson 2006; Reid *et al.* 2011; Powers *et al.* 2013). As these predators (most notably, white sharks) show age-related changes in prey preference, a shift towards smaller individuals could lead to different behavioural patterns of shark populations in a particular region. For example, more individuals may frequent in-shore waters to chase smaller prey, as opposed to focusing on larger marine mammals off-shore, thus increasing the overlap in in-shore water usage between sharks and humans, and the potential for negative interaction.

Why sharks are important

When faced with the conundrum of human–wildlife conflict, especially the dilemma of protecting human life at the expense of another animal, there is always the question of why should we even bother trying to preserve the life of the other animal. The valuation of one animal's life over that of another is clearly a subjective matter, and everyone will have a different opinion based on their own cultural and personal beliefs. However, leaving broad conservational and ethical reasoning aside, there are several reasons why we really should reconsider killing sharks.

The role of sharks in successful ecosystems

Sharks have an important role in the ecosystems they inhabit. However, the exact role they play varies by species, region, age/size, behaviour and/or sex. The common thought that sharks, as a whole, are apex predators is incorrect. In fact, very few are. Apex predators occupy the highest trophic level and have substantial top-down effects on prey demography and the structure and productivity of their ecosystem through direct (consumptive) regulation of prey dynamics and indirect (non-consumptive) modification of prey behaviour. Direct predation results in the prevention of drastically increased intermediate predator numbers (mesopredatory release). Indirectly, the presence of predators in an ecosystem enhances biodiversity through modification of prey species' foraging behaviours, spatial use and interactions. Some of the larger shark species (sometimes referred to as the 'great sharks') would certainly be classified as apex, or top-order, predators but, when considering diet and tropic ecology, most sharks are more accurately mesopredators. Overall, the breadth and diversity of the functions that sharks play in their respective ecosystems, and at varying levels of the food chain, would be enormous. The removal of apex predators from an environment would have the most obvious and significant effects on the ecosystem because ecologically redundant species would not be as available to mitigate and lessen the impact. Resulting tropic cascades would be likely, with effects propagated throughout the ecosystem. Although not as recognisable, the removal of mesopredators would still be detrimental to an ecosystem, as these species also play a variety of important roles in the environment, including maintaining the structure, function and reliability of the ecosystem.

Indirect effects of predator presence are broad-reaching, and affect everything from flora and fauna composition, productivity, nutrient cycling and (transient) water quality parameters, trophic transfer efficiency and energy flux (Schmitz *et al.* 2008). For example, tiger sharks are seasonal migrants to Shark Bay (Western Australia), and their seasonal presence has a variety of implications for resident species such as green turtles (*Chelonia mydas*) and dugongs (*Dugong dugon*). During periods of relative safety, when predatory sharks are in low numbers, dugongs risk the reduced visibility created by the excavation of preferred and more energetically profitable tropical seagrass rhizomes. However, when sharks are present in greater numbers, dugongs forgo these optimal feeding opportunities and instead utilise the safer foraging method of cropping temperate seagrass, which allows for regular visual scanning. Thus, through the reduction in dugong disruption to seagrass structure and composition, tiger sharks indirectly affect the overall local composition and structure of seagrass assemblages and the benthic community (Wirsing *et al.* 2007).

Sharks as ocean cleaners

Although commonly alluded to, the behaviour of sharks preying on sick, injured and old animals (and thus removing them from the population) is not often reported in scientific publications. It is a logical concept, solely on the basis of energy expenditure, as suboptimal animals would theoretically be easier to catch and consume. This concept also suggests that, in a similar fashion, sharks (and other predators) encourage the evolution of larger, stronger animals, as those are the ones more capable of avoiding predation, and surviving to reproduce and pass on their genes. However, unhealthy animals in poor body condition may not provide the same energetic reward as healthy animals in good body condition. Some theories state that predators could, in fact, contribute to the spread of disease, for example by parasite distribution through feeding or defecation, the removal of animals from the breeding population that have developed immunity and recovered from infection, or the alteration of normal behaviour in the presence of a predator to one that increases host susceptibility to parasitism (Duffy *et al.* 2011). This is not necessarily suggested for sharks, but the concept of sharks eating only sick, injured or old prey and 'cleaning' the oceans is a sweeping statement. It is also difficult to prove; this is most likely why it is lacking from scientific studies. Interestingly, though, the level of risk that potential prey take in the presence of sharks can be directly correlated with health and body condition. Green turtles in poor body condition were found to utilise more profitable but high-risk habitats for foraging, despite the threat of tiger sharks, whereas turtles in good body condition preferred safe but less profitable microhabitats (Heithaus *et al.* 2007a). Thus, suboptimally conditioned animals were more exposed to sharks.

How sharks may benefit human health

Sharks are incredible animals and, while they are often the focus of attention for all the wrong reasons, they also teach us some really incredible things. Sharks are survivors in more than just their predatory capacity. As mentioned, sharks have been in existence for a long, long time, so it is not surprising that they have evolved various mechanisms that allow them to survive conditions that many other species would not. Indeed, shark antibodies are currently being studied by the US Defence Threat Reduction Agency and Naval Research Laboratory for the development of diagnostic and therapeutic tools for protection against chemical and biological threats (DVIDS 2016). We have learned and will continue to learn from sharks, and it will be very exciting to see what new health insights we gain from sharks in the coming years.

Anoxia tolerance

My husband used to be a shark researcher, and he focused on learning about shark stress markers and the anoxia tolerance of sharks. Anoxia (the complete absence of oxygen) is not something that would be commonly thought of in the context of sharks, but it is highly relevant to human medicine. Anoxia or hypoxia (the milder situation of reduced oxygen) result when oxygen is unavailable to cells or tissue within the body. Anoxia is generally a catastrophic medical situation for humans, and some degree of hypoxia is often encountered in patients suffering from heart attacks or respiratory arrest, strokes, brain tumours, low blood pressure, severe asthma or anaphylaxis, choking and newborns

Clint Chapman taking a blood sample from a brown banded bamboo shark (*Chiloscyllium punctatum*) in order to study the shark's physiological response to anoxia stress. Photo by Blake Chapman.

facing complications during birth. Hypoxia-ischemia, for example, is a major cause of neonatal death and long-term disability in newborns (Lai and Yang 2011). Cells within the human brain begin to die within four minutes of severe hypoxia, and permanent brain damage occurs after approximately five minutes. Death ensues shortly after. However, several vertebrate species have developed coping mechanisms for surviving (and functioning) in periodically hypoxic environments. Included in this very select group are the epaulette shark (*Hemiscyllium ocellatum*) and the brown banded bamboo shark (*Chiloscyllium punctatum*), which can survive completely anoxic conditions at ambient (subtropical) water temperatures of 25°C for two hours and one hour, respectively (Chapman *et al.* 2011). This is a remarkable adaptation, especially given the temperature range at which they do this. It is expected that these species have developed the relevant coping mechanisms to functionally exploit a physiologically challenging niche that prey may be able to survive in, but not necessarily actively function in. Further investigation into the physiological mechanisms used by these species may lead to advanced therapy and counteractive measures for hypoxically challenged humans.

Immunology

While working in animal health at the aquarium, I spent many hours observing and documenting injuries to the animals. In this regard, the sharks did not disappoint. Injuries arose from both natural and artificial causes, including mating, predation and misadventure with exhibit infrastructure, and ranged from minor to fatal. I was always amazed at the healing ability of the sharks, in particular. One case I worked closely on involved an adult blacktip reef shark (*Carcharhinus melanopterus*) that had clearly had a negative interaction with a larger shark. Its tail had been broken and sat at an unnatural angle. One of its pectoral fins was sliced nearly in half and there were accompanying bite wounds to the torso and tail. This appeared to be a significant injury, which needed substantial veterinary treatment, pain relief and antibiotics. However, despite the seemingly significant injuries, the relatively small, confined exhibit space that the animal was in, and many hours of effort by an expert team of divers and animal handlers, the shark could not be caught. Blacktip reef sharks are one of the more skittish aquarium species and this animal could not be target-fed like many of the other sharks kept, so there was no way of getting injectable or oral medication into it. Yet it continued to thrive. At no time did it show any signs of decreased body condition or spreading infection. Even the deepest flesh wounds healed within weeks and the broken tail reset itself in time as well, to the point where no deformity could be distinguished. Although the fracture was unique, flesh wounds to the sharks were common. Every time, they healed with remarkable rapidity. Again, thinking about the biology and behaviour of sharks, this is not completely surprising. These animals are predators, and often prey on large, well-protected animals that could cause damage in several different ways, be it through claws, teeth, spines or shells. Furthermore, mating behaviour in many species of sharks is often highly aggressive and traumatic, leading to extensive wounds to both females and males. Therefore, the ability to quickly recover would be highly advantageous for survival and successful reproduction.

The healing ability of sharks has not gone unnoticed by human medicine, and in fact, there has been a boom in recent years in the study of shark immunology. A naturally occurring aminosterol originally derived from the dogfish (*Squalus* spp.) liver has been shown to increase the regeneration rate of amputations in fish (including bone and tissue) by two to three times and increased survival and improved heart function in mice with induced heart attacks (Smith *et al.* 2017). Compounds obtained from elasmobranch tissue have shown cytotoxic activity against tumour cells, antiangiogenic activity and antibiotic activity. Although elasmobranchs may well represent the most primitive group of vertebrates that contain the full suite of components necessary for an adaptive immune system as higher vertebrates, immunity in this group differs (Walsh *et al.* 2006). They lack bone marrow (where B-cells, which produce antibodies, are typically developed), and their types of immunoglobins and genomic organisation additionally set them apart from other vertebrates (Marra *et al.* 2017). For some time, the (incorrect) belief that sharks do not get cancer spruiked further interest in elasmobranch immunology. Various elasmobranch-derived substances have been, and are currently, under clinical trial for various lung, colorectal, kidney and breast cancers, macular degeneration and even hair thinning. Recently, several shark genes have been identified that play important roles in various aspects of immunity, which could be related to gene overexpression in certain cancers and eliminating tumour-associated macrophages, among other things (Marra *et al.* 2017). Synthetically prepared squalamine (a shark-derived aminosterol compound with broad spectrum antibiotic/antimicrobial properties) was found to not only slow down but also remove the toxicity of the protein aggregations associated with Parkinson's disease in an animal model (Perni *et al.* 2017). Further testing will be done and, if successful, may lead to a drug effective in treating or relieving symptoms that the many millions of people living with this disease face. Yet another drug inspired by an antibody in the blood of sharks has been developed, and initial testing successfully targeted fibrosis-causing cells that can lead to the debilitating and currently untreatable condition of idiopathic pulmonary fibrosis (Salamastrakis 2017). This condition causes persistent and progressive scarring and build-up in the lungs, reducing and eventually preventing oxygen and carbon dioxide exchange with the blood, at which point it is fatal. These are only some of the recent human health advances stemming from shark immunology, and it is expected that more human therapeutic discoveries will eventuate as we learn more and more about sharks.

Shark products are heavily marketed for non-fatal degenerative conditions in both human and animal pharmaceuticals. Shark liver oil contains a high concentration of polyunsaturated fatty acids, which are important for brain development and the prevention or reduction of mental health disorders (Ozyilmaz and Oksuz 2015). It is also marketed for optimising the health and function of the immune system, respiratory ailments, skin conditions, antioxidant activity and improvements to fertility and overall well-being. Shark liver oil may be used along with normal cancer drugs to help alleviate radiation illness and increase white blood cell counts during treatment and to help fight cancers such as leukaemia. Squalene, which is found in shark liver oil, and squalane, which is derived from squalene, are considered to be of use in wound healing and the prevention of bacterial multiplication. However, the efficacy of shark-derived pharmaceuticals is often proven to be equivocal.

Shark cartilage is promoted as a natural, non-toxic therapy for osteoarthritis and rheumatoid arthritis in humans due to its anti-inflammatory properties. However, it is more widely prescribed as a supplement for the treatment of arthritis and joint injuries in dogs. It's also used in the treatment of canine cancers, as chemicals within the cartilage are thought to inhibit the formation of blood vessels, preventing required nutrients from reaching tumours and halting their growth. Large doses of shark cartilage are required for successful therapy results.

As we are only just starting to identify and isolate the specific beneficial agents in shark products, few synthetically derived alternatives exist (Adlington *et al.* 2016). Therefore, the production of these supplements still requires killing sharks. The three species at the core of the shark liver oil industry (which provides controversial human health benefits, at best) – the leafscale gulper shark (*Centrophorus squamosus*), the spiny dogfish (*Squalus acanthias*) and the basking shark (*Cetorhinus maximus*) – are all listed as vulnerable (with decreasing population trends) on the IUCN Red List of Threatened Species (IUCN 2016).

Conclusion

Despite the many fascinating characteristics of sharks, the thing that people most often associate these animals with is their potential for negative interactions with humans: shark attacks. Shark attacks have become a highly mediated, sensationalised topic and attract a great amount of interest from a wide variety of people. Not surprisingly, opinions on shark attacks can be very strong and greatly divided. Although shark attacks are extremely rare, they can be highly traumatic to those affected and the ripples of their effects are often widely felt. Any attempts to downplay the potential severity and trauma of these events would not only be inappropriate and misinformed, but also highly disrespectful to the individuals who have been through these situations. As shark bite incidence increases globally, we have a growing responsibility to address the topic in a clear, unbiased and comprehensive manner. To more completely understand the subject of shark attacks and identify ways to mitigate the risk of such events, we need to gain greater insight into shark biology, human psychology and the influence of our media.

2

Shark biology and basics

Although incredibly interesting, the biology, physiology and ecology of sharks are topics that are far too large and broad to be extensively covered here. However, the concepts of shark attack and shark attack mitigation cannot be comprehensively discussed without at least an abridged understanding of certain relevant aspects of these subjects. Therefore, this chapter provides a little bit of background information on select areas of shark biology and physiology that are most relevant to the topic of shark attack and the material discussed throughout the remainder of the book.

Attack-relevant anatomy

In response to the word 'shark', the most common mental image would be that of a large, sleek, streamlined body, deep black eyes and possibly large exposed teeth. And while this description may be accurate for some of the more stereotypical sharks, these characteristics are far from defining in this group of animals. While there are some common and characteristic attributes among sharks, there is a greater degree of diversity between species.

While many sharks do have streamlined, torpedo-like bodies, there are also many examples of sharks that are round, elongate or even flat. Wobbegong sharks (*Orectolobus* spp.), for example, look like they have been run over by a steamroller. Yet, despite their ungainly and seemingly inefficient body shape, they are master ambush predators and have been responsible for 32 unprovoked bites on humans in recorded history – a decent number (ISAF 2017). The proportions of sharks' bodies and fins have been described as flawlessly adapted to their environments (Sudo *et al.* 2002). Fin size and placement and tail morphology are closely linked to species behaviour. Certain body types facilitate greater swimming speeds, whereas others favour reduced energy expenditure through increased streamlining and reduced drag or manoeuvrability. The pectoral fins are critical for locomotion, as they control body position and facilitate rotation. It is the fins of sharks that allow them to perform many of the characteristic manoeuvres that we often think about – the predatory breaching and aerial acrobatics of white sharks (*Carcharodon carcharias*) and even the lovable 'walking' behaviour of certain benthic sharks.

The speed and hydrodynamic capabilities of sharks are not only due to body form and musculature, but also to characteristics of their placoid scales. These collagen-embedded scales are known as dermal denticles and act to break the barrier between the skin and water, minimising turbulence and decreasing drag. Combined, these attributes allow sharks to reach burst speeds of up to 70 km h^{-1}, as seen in the shortfin mako shark (*Isurus oxyrinchus*) (Diez *et al.* 2015). They also protect sharks from predators and ectoparasites and reduce mechanical abrasion. Certain species of sharks may even use their scales to anchor pieces of food for biting and tearing (Southall and Sims 2003). The tooth-like scales are essentially a modern remnant of ancestral external bony armour. They cover the entire body, including the fins, the retractable protective nictitating membrane that covers the eyes of some species, and the gills (Meyer and Seegers 2012). Although denticle shape varies significantly between species (and over different parts of the body), they typically end in sharp points at the trailing edge. As a result, shark skin is remarkably smooth in the head-to-tail direction, yet highly abrasive and incredibly destructive from tail to head.

The pointy end: shark jaws and teeth

My first shark bite occurred, ironically, while scolding a research assistant on their improper shark-handling technique. This person had picked up a small blacktip reef shark (*Carcharhinus melanopterus*) from a holding tank and was holding it by the tail. The shark was writhing around, snake-like, with its head and torso unrestrained. I rushed over to gain control of the animal before either the shark or the person was injured. Despite aiming to grab the shark just behind the gills with

The author, following her 'shark bite' as a novice research student. Photo by Clint Chapman.

both hands, I missed. I can't remember feeling anything at the time, but I knew I had been bitten. The shark had cut the top corner of my thumb and there were a few minor scratches in my fingernail. That was the meagre extent of it. The photos that were taken later that evening show me with my thumb bandaged with a single Band-Aid, pretending to be injured. But mostly, I was grinning from ear to ear. For a novice shark enthusiast with an admittedly pitiful pain threshold, this was sufficient to let me tick 'being bitten by a shark' off my bucket list. It also led to my interest in shark jaws and teeth.

In general, shark jaws (and their related components) are structurally simple, composed of just 10 cartilaginous elements, compared to bony fish, which have more than 60 bones (Motta 2004). Sharks may capture prey by ramming, biting, sucking or filtering. Depending on the size of the shark and the prey, the process of biting and capturing prey may occur through a multi-staged sequence, or prey can be passed directly through the mouth. Bites can capture,

A white shark (*Carcharodon carcharias*) displaying its jaw gape. Photo by Denice Askebrink.

manipulate or process food items. Capture bites generally occur very rapidly, in the order of 0.1–0.4 s.

Sharks are renowned for their teeth, and one of the first things that people often learn about sharks is that they regularly replace their teeth. Shark teeth generally develop in rows. Most sharks have 20–30 rows of teeth at any given time, although some species have up to 300. The size and shape of shark teeth are species-specific, and nearly all species display monognathic heterodonty, or significant morphological variation among teeth in different parts of the same jaw. Many species also display sexual and/or age-related variation in tooth morphology. Jaw and tooth morphology can be so distinct to a species that the pattern of bite imprints and tooth fragments in objects recovered from shark bites can often be used to identify the species. Shark tooth shape and size is often considered to be directly related to functionality and can be a primary predictor of feeding behaviour. Smooth, slender tooth morphologies, as in shortfin mako sharks (*Isurus oxyrinchus*), which feed primarily on fish and cephalopods, are best suited for piercing, while the serrated teeth of white sharks, which have a far more variable diet, including larger prey items such as marine mammals, are best suited for slicing and cutting and make better use of applied bite force (Stillwell and Kohler 1982; Frazzetta 1988). Tiger shark (*Galeocerdo cuvier*) teeth are strikingly asymmetrical (and easily distinguishable) and are multifunctional. The anterior and posterior tooth margins are coarsely serrated and the curvature of tooth margins converges towards a distinct notch on the outside edge of the crown. Although individually asymmetrical, teeth are arranged in mirror-image on either side of the jaw, so when the shark swings its head from side to side, gripped prey are subject to both aspects of the tooth margin. The thin notch increases stress to the area and is thought to facilitate the cutting of tougher materials, such as ligaments, tendons and collagen bundles (Motta 2004). Feeding isn't the only function of sharks' teeth, though, as they also rely heavily on them for reproductive purposes. The males of many species firmly implant their teeth into the tough skin of females to lock onto them during copulation.

Tiger sharks (*Galeocerdo cuvier*) have highly asymmetrical teeth, which may just be visible during normal swimming behaviour. Photo by Denice Askebrink.

The bite force of sharks can be related to trophic ecology and fitness. A greater bite force allows for the consumption of a wider range of prey types, and therefore represents a selective advantage. When considering typical prey items of larger sharks, greater bite force could mean the difference between being able to consume a marine mammal, with stiffer and more resistant skin and bones and larger skeletons, and smaller, softer, less energetically rewarding fish. Of the species studied, the great hammerhead (*Sphyrna mokarran*), bull shark (*Carcharhinus leucas*) and white shark have the greatest bite forces. Bull sharks have the greatest calculated size-specific anterior bite force (and one of the highest among extant vertebrates) (Habegger *et al.* 2012). The bite force of a 285 cm (193 kg) bull shark was estimated to be ~5900 N at the back of the jaw and 2128 N at the front. When considering relevant prey items, this force represents a gross over-design of the feeding mechanism. However, the over-design may be related to the shark's feeding ecology, which includes murky water, and the need to bite and hold onto prey, negating further energy expenditure involved with relocating or recapturing an item not contained during the initial bite. Thus, the bite force could allow an extremely forceful bite-and-grip behaviour. Over-specified bite forces could also be necessary for species that bite and grip prey and then use head-shaking to rip or dismember the prey. Although not currently quantified, tiger sharks are likely to have extremely high bite forces based on their prey preferences and feeding behaviour.

Physiology and habitat

Being aerobic organisms just like us, sharks rely on oxygen to survive. Except that, in sharks, oxygen is harvested from water passing over the gills. Sharks do this either through the process of buccal pumping (actively suctioning water through the mouth and over the gills) or ram ventilating (holding the mouth open to allow water to flow through the gills, either as the shark moves forward or through water movement). Obligate ram ventilators must continue to move to fill their oxygen demand, which consequently increases the energy needs of these species. The variations in respiration strategies are most apparent during resting behaviours; whitetip reef sharks (*Triaenodon obesus*), for example, are buccal pumpers and are often observed lying on the substrate. Buccal pumping sharks are often provoked by divers, as they can be easily approachable. Grey nurse sharks (*Carcharias taurus*) are ram ventilators, and may hover relatively stationary in a current with their mouths open. Sandbar sharks (*Carcharhinus plumbeus*), on the other hand, as obligate ram ventilators, will always be moving. The variations in the respiratory requirements of different species have major ramifications in terms of their survival ability if they are captured in nets or on lines that significantly restrict swimming.

The northernmost recorded shark attack occurred at Vityaz, Russia (42°N) and the southernmost attack appears to have occurred at Campbell Island (52.5°S), a sub-Antarctic island around 700 km south of New Zealand (Brenneka 2009; ISAF 2017). Both attacks have been attributed to white sharks. The body temperature of most fish is closely linked to water temperature, as aerobically produced heat is lost through the gills and body surface. However, Lamnid sharks (and, interestingly, convergently evolved tuna), have developed a highly evolved, specialised physiological adaptation that allows them to maintain their body temperature above water temperature. This adaptation, which utilises counter-current heat exchangers and is most notably found in salmon sharks (*Lamna ditropis*), shortfin mako sharks, white sharks and common thresher sharks (*Alopias vulpinus*), is intrinsically linked to the animal's ecology and energetics. In sharks, the anatomical adaptations that allow for this, called *retia mirabilia*, can be found in the skull near the eye, locomotor musculature and abdominal organs, regulating the temperature of the brain and eye, red muscle, liver and kidney (Bernal *et al.* 2001; Lowe and Goldman 2001). To further conserve body heat, the red muscle of these species is located close to the spine, as opposed to near the body walls (as in ectothermic fish). Body temperature in endothermic sharks can be further regulated through vascular shunts and sinuses to bypass the *rete*, allowing for precise regulation of heat retention (e.g. when in warmer water), and overall temperature gain and loss. The remarkable efficiency of this system allows for an elevation of body core temperature of up to 21.2°C (in salmon sharks) above ambient water temperature (Goldman *et al.* 2004).

My friend and I were swimming in the ocean to an island that's about a mile off-shore. We were training for the 9.5 mile Maui Channel Swim, an annual swim that happens in Hawaii that goes from Lanai to Maui. I regularly surfed and swam here in Hawaii, and this was a swim that I had done probably 30 or 40 times previously, and I know people who do it almost every day. We have all sorts of different sharks in Hawaii, and I've seen many sharks at various points, but I've never been concerned with identifying which shark was which and I've never been really afraid of them. None of the sharks I had seen were ever aggressive, and I'd never had a scare before, so I wasn't worried about them. On the day, it was clear outside, but the water was a little murky. I'd swum in murky water before, but I guess murky water is a big thing.

We had swum to the island, and were on our way back when we got caught in a rip current and were separated. I reached the open channel first, with my friend still about a hundred yards away, so I slowed down to wait for him to catch up. I didn't see the shark at all. Afterwards, a friend told me that sharks only attack by ambush, so if I had seen it, it may not have attacked me, but I didn't see it. And I didn't feel foreboding, or a sense of anything, I was completely surprised. It grabbed both of my legs, and at first I thought just for a second that it was my friend, that he had caught up with me and was just messing with me, but then I felt the horrific pain. Later, they determined that it was a tiger shark, over 12 feet long. The shark just kept chomping. When I looked back, I mostly just saw the water thrashing. I didn't see the animal, as such, at the time, I just saw a big shark. I reached back, and I was punching him with my right hand as much as I could, but that didn't do anything. I even got him right on the snout. People often say that if you punch a shark right on the nose you can deter it, but in this case, it did nothing.

The shark then started driving me down into the water. I had goggles on, so I could see everything. And I know it's weird to say this, but I was amazingly calm. I pulled up to grab a quick breath of air. I'm a surfer, so I know that when you grab air, you grab a deep lungful, so that's what I did. But the shark then just kept pushing me deeper and deeper. I kept punching him the whole time. And maybe it was because I was really freaked out, or whatever, but I felt an incredible amount of pain right at the beginning, but then didn't feel any pain at all. But to be fair, the shark had stripped all of the meat off of my legs, they were both just bone, so it probably took all the nerves. And without any nerves, there wouldn't be any feelings of pain.

Eventually, I felt my ears pop. At this point, I knew I was at least 10 to 15 feet under the water, because I scuba dive, and I know that that is the point where my ears pop. But we still just kept going deeper. I could feel the air in my lungs collapse, and I just kept thinking 'I'm really in trouble now'. The shark kept its grip on my legs the whole time; it just kept holding on and pushing deeper, and then my right leg broke, or just came off, and I was able to turn around and face the shark. That was the first time I really saw him.

I do research into eyes for a living, and from this research I know how to efficiently and effectively take the eye out of an animal. I know that if you push really hard in the medial aspect of the eye, that you can get your finger behind the eye, and it will pop right out. While my professional experience had mostly been with pigs and rabbits, I knew that this was my best chance, so that's exactly what I did. Tiger sharks have a nictitating membrane, just like rabbits and pigs, so I used one finger to pull up the membrane, and I just pushed really hard with another. I used my index finger to get right behind the eyeball at that medial aspect, and I pulled and ripped the shark's eye right out. And luckily it let go of me.

I surfaced as quickly as possible, but I've got to say, it felt like it took forever. I know it probably wasn't anytime at all, but you know how time gets distorted in your brain, and it felt like I was so deep. But I finally got there, and I still had the shark's eye in my hand. I looked around, but didn't see anyone except for my friend, who was still 50 yards away, or maybe even 100, or 150 yards. I just started yelling for help. I was so worried that the shark would come back, but I knew I had to move and do something, so I laid on my back and started to backstroke towards shore. We were still about half a mile out to sea at that point. I know this now because my friend had a GPS watch on, and we were able to later look at the results, so we have an idea of exactly where the attack happened. I was still on my back, yelling and screaming, while

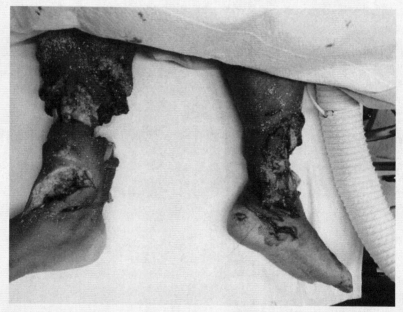

Tony Lee's leg injuries following an attack by a 3.7–4.3 m tiger shark (*Galeocerdo cuvier*). Photo supplied by Tony Lee.

heading towards the shore, and out of nowhere, a father and son on two out-riggers came up to me. When I had looked around previously, I hadn't seen them, but I had been paddling for about five minutes, and they just appeared out of nowhere. The father came up to me and I crawled on his boat. I remem-ber getting on the boat. It was an outrigger, so sort of like a thin canoe with an arm, and a race boat. They're very unstable, and we flipped a couple of times. The man used his rash guard and tied a tourniquet on my leg and I laid on the boat with the one leg I still had and braced it against the pontoon. And we finally got to the shore. Meanwhile, the son had gone to get my friend, and that's when the boy saw the shark. He identified it as a tiger shark and said it was as long as his boat. Luckily my friend and the boy beat us to the shore, so by the time we got there, they had already called the ambulance and it was only about five minutes away.

When the ambulance arrived, I was moved into it. Surprisingly I never blacked out, and I remember everything. And I have specific memories of the paramedic working on me. By that time I was completely white, and I could barely breathe, because I had very little blood left in my body. The paramedic was trying to cannulate me, to get a needle in my arm, but the veins in my arm had all collapsed. We were in the ambulance with the door still open, and we were just sitting there. The paramedic knew he had to get this done before we could start moving. But by this time, there was a big crowd that had gathered, and everyone was like 'get moving, come on, get moving', but he knew that he had to get the needle in before the car could move. I remember looking down at him and he was just sweating. It's hard enough to find a vein when every-thing is right, but when everyone is yelling at you to 'GO, GO, GO!', and when my veins were completely collapsed, and when he would have known that I probably only had about 10–15 minutes left of life, I mean I've got to hand it to this guy. I was just watching him, and I remember thinking to myself 'Man, I feel bad for this guy, I hope I don't die on him because he'd feel really bad about not being able to get that needle in.' He finally got the needle in, and started injecting fluid into me. The ambulance didn't have blood on board, but they had saline, and they started putting that in. You need blood, but you also just need volume in your veins, because if you don't have volume, your heart can go into cardiac arrest. And unfortunately, I knew that, so I was sitting there thinking, 'I'm going to go into cardiac arrest if you don't get some fluids into me real quick.' He started pumping the fluids in as fast as he could, closed the door, and we rushed to the hospital. There was a monitor in the ambulance, and I could see my vitals, and the paramedic just kept saying, 'Hey, don't look at the monitor, don't look at the vitals', but I couldn't help it because they just kept dropping. And I couldn't help but think, 'Oh, yeah, that's not good.' My oxygen level had gotten really low, like medical school textbook low and beyond what people die at low, and I was thinking, 'Oh, my God, this is horri-ble'. Finally, he just turned the screen away from me, because he didn't want me freaking myself out. But we finally got to the hospital, and they filled me full of blood, and somehow I lived.

I have teeth marks all over the sides of my legs. Sometime after the attack, a guy from National Geographic came out and interviewed me and he measured the size of the teeth marks and how far apart they were, and he estimated the shark to have been 12–14 feet. And that's also what the boy on the outrigger estimated, so the two numbers match up.

I just had the year anniversary of my attack about a week ago. I still have specific memories, and I remember everything. I suffered a little bit from PTSD, particularly in the form of bad dreams for a little while after the attack. A while after the shark attack, I went to the swimming pool and I really freaked out when I was in the water. I could hear the shark, it was the weirdest thing. The thrashing, the sound of that violent thrashing, I could still hear it. I've been back in the ocean since the attack. I've been surfing, and I've done some long ocean swims. I haven't been back to the same place, but now when I swim in the ocean, I constantly look behind me.

I never wished for retaliation on the animal, and I definitely don't think that culling is a good idea. You could mess up the ecosystem. Sharks have a place and if you take all the sharks, then the jellyfish – or some other population – would become out of control. I know they do retaliatory culls sometimes. I've since met a lot of other shark attack survivors, and one guy was telling me that back in the 1980s or '90s, when he got attacked, they did a cull and killed dozens and dozens of sharks. And he seemed happy about it, but I was thinking, 'Man, I really don't want them to go out there and start slaughtering sharks just for me.' But other mitigation measures, like warning people – yeah, that would be good. If there is a shark in the area, I'd definitely want to be warned. And from now on, when I'm swimming in the ocean, I'm going to stay in a group, because from what I've heard, sharks normally don't attack groups of people.

Sharks have to eat, and in this case, I think the shark was unfortunately trying to eat me, or at least my legs. I was a meal. I can't say that for certain, but it kept attacking. I had heard before that sharks attack due to accidental mis-identification, but this wasn't the case here. I also spend a lot of time in the San Francisco Bay area and we have great whites there. And every now and then, we have a shark attack. But with them, they normally take one bite and then go. The minute they take a bite and realise the person isn't a seal, they let go. But the tiger shark definitely did not, and I was definitely doing my best to deter it, punching and punching and punching. I have pictures of my knuckles while in the hospital three or four days after the attack, and my knuckles were still completely raw and bloody. I had punched that thing until the bones were showing in my knuckles. I wasn't just giving it love taps, I was really punching it hard. But it didn't dissuade the shark at all. Tiger sharks in the region have their babies around late September or early October, so it was probably a female shark, either heavily pregnant, or who had just given birth, and probably starving. So I believe it was just doing what it needed to do to survive.

The injuries I sustained resulted in one of my legs being amputated and the other reattached. I have a lot of trouble walking and getting around now, and I was upset about that. It's sad, because I used to run, surf, bike and swim. I can still bike, kind of, but I have to have a special attachment. And walking – I can't walk for more than 10 minutes. Things are a lot harder now. Surfing is really hard; it's really hard to balance on the board. And unfortunately, it's not the prosthetic that the problem, it's more the reattached leg that's the problem. It's pretty gruesome looking. All the meat was gone, so they had to take a couple of muscles out of my back and attach them to the leg. And they took tendons from a cadaver, and used metal rods to reconstruct the whole leg. It's all kind of a mess down there. I had to undergo another surgery on that leg about two months ago, and they told me that I may lose that leg as well, so I had to come to grips with losing both legs. They were able to save it then, but I may still have to have it amputated at some point because it's still so painful right now, and it's still so difficult to walk on. It may end up being easier to just take it off. But, I'm getting around now. And even at 45 years old, with just one reconstructed leg,

Tony Lee getting on with his active lifestyle in the year following his attack, despite his debilitating injuries and amputation. Tony continues to hike, rock climb, surf, swim and cycle, and has competed in organised distance ocean swims and cycling events. Photo supplied by Tony Lee.

I've finished a 25 mile bike race, and a mile and a quarter long swim, so I'm getting back out there and trying to not let this stop me.
  I know a couple of other guys who have been attacked. Some are happy to talk to people about the attack, others aren't. Some of them are really bitter and angry about it still, even though the attack may have happened 20 years ago. It's a small fraternity of people that have been attacked by sharks, but it's been really helpful to have had other people to talk to. Another guy was attacked around the same time in Hawaii, and he was in the hospital at the same time I was, so they just put us together. I was put in contact with others through friends. I don't know of any organisation that connects shark attack victims, but maybe there should be, because it was so helpful. I also had people coming into the hospital while I was there – other survivors. They just came in. There was such great support, and that continued support from other survivors gave me the inspiration I needed to get on with my life.

Tony Lee

Sharks are best known for their coastal interactions with humans; however, they occupy nearly every aquatic environment explored. While most sharks occupy marine (salt water) environments, some species also occupy freshwater habitats. Bull sharks have the unique ability to move freely between marine and freshwater environments, and are therefore classified as euryhaline (Pillans *et al.* 2005). While bull sharks were long thought to be the only shark species capable of such movements, the more mysterious river sharks (*Glyphis* spp.) have recently been found to have the capacity to move across salinity gradients (Li *et al.* 2015). The wide habitat range of these sharks notably extends the area of overlap with those commonly used by humans, to include rivers and even lakes (if they connect, even temporarily, with the ocean).

Oceanic whitetips (*Carcharhinus longimanus*) are considered to be one of the most dangerous species of sharks to humans. However, the number of reported attacks does not necessarily reflect that danger (ISAF 2017) because the largely off-shore, oceanic lifestyle of these sharks means that they overlap with humans far less frequently than coastal species. Oceanic whitetips are mainly a threat to people who have experienced ship or plane wrecks, and it is likely that they are responsible for unrecorded oceanic human fatalities.

Bull sharks (*Carcharhinus leucas*) are capable of inhabiting a range of environments, from salt to freshwater, and are often reported in shallow waters. Photo by Tyrone Canning.

Cognition and learning

Wobbegongs are to sharks as the stereotypical Hollywood jock who has had one too many knocks to the head is to the human race. They are the 'bite first, ask questions later' species with what I consider to be a 'flat shark' complex. We handfed the large sharks at the aquarium, and I fondly remember carrying out many feeds with a wobbegong firmly attached to my foot. Luckily, their teeth aren't long enough to fully penetrate dive boots, but having a 2 m wobbegong attached to your foot was like being tethered to one spot with a cinder block, which, needless to say, wasn't ideal when we were also feeding swift and agile 2–3 m whalers and grey nurse sharks that did have more formidable teeth! In another example of the intelligence of these sharks, a colleague told me his experience about catching a wild wobbegong. He trapped the shark in a gill net then scooped it into a hand net to bring it to shore so he could take measurements and a blood sample. Wobbegongs are famous for being extremely flexible and lightning-quick (tip: don't ever pull a wobbegong's tail!) but as previously mentioned, not for their mental capacity. This wobbegong was clearly distressed in the hand net and decided to bite something in an attempt to change the situation. Unfortunately, the only thing available to bite was its own tail. By the time my colleague got the shark to the beach, the shark was in an inseparable ring. No matter what he did, he could not get the shark to release its grip. Eventually, he had to release it without getting any of the information he needed. He went back the next day out of curiosity and the shark was still there, right where he released it, and still biting onto its own tail.

Wobbegongs, however, should not be used as a model for shark intelligence overall. Sharks have widespread variation in brain size, organisation and complexity, but there are patterns that link species with similar diet, habitat and social behaviours. Sharks were the first to exhibit the conserved vertebrate brain model, which includes the forebrain (telencephalon and diencephalon), midbrain (mesencephalon) and hindbrain (cerebellum and medulla oblongata). Although historically, and perhaps still thought by many (especially after reading the accounts of wobbegongs) to be mentally inept, sharks have been shown to possess brain size to body ratios comparable to those of birds and mammals (Yopak 2012). Species that occupy complex reef or oceanic habitats tend to have larger, more developed and foliated brains, likely allowing them to be more strategic, agile and flexible in hunting and possessing a greater capacity for complex learning. They also may have increased social intelligence.

Sharks have a demonstrated capacity for learning, with a rapidity in line with that of other vertebrates, and a memory span of at least several months. The ability to learn and remember is generally considered most important for functions like foraging (search, capture and manipulation), travelling (orientation, navigation and migration) and social interactions (with mates or conspecifics). This ability provides flexibility in navigating a variable environment on the basis of previous experience, and sharks have been shown to become more accurate and efficient in their movement between landmarks with experience (Guttridge *et al.* 2009; Papastamatiou *et al.* 2011; Gallagher *et al.* 2015). Laboratory studies have also proven that positive reinforcement (e.g. through a food reward) can markedly and rapidly alter shark foraging behaviour (Kimber *et al.*

2014). Ultimately, the intelligence of these animals, and their capacity for learning, should not be taken lightly.

Sensory biology

As compared to the five sensory modalities of humans, sharks possess at least six major sensory systems that allow them to detect prey and conspecifics, avoid predators and obstacles, and navigate through their (often expansive) environment. These modalities include vision, hearing, touch (through somatosensation and mechanoreception), electroreception, smell and taste. The efficacy of sensory systems is evaluated on the basis of two criteria – sensitivity and acuity. Sensitivity relates to the minimum stimulus needed for detection, whereas acuity determines a system's ability to discriminate stimulus characteristics. The sensitivity of shark sensory systems varies widely between species, environment and developmental stage. However, generally, taste and somatosensory capabilities rely on the stimulus being in contact with the sensory receptors, electroreception and mechanoreception can be sensed in the range of centimetres to metres, vision can extend to 100 m and smell and hearing can be stimulated from distances of hundreds or even thousands of metres (Collin 2012). Although fascinating, and important for many reasons, our understanding of the sensory capabilities in sharks is still limited. While laboratory behavioural analyses of live animals and anatomical and physiological assessment of sensory tissue have provided a great deal of insight, our inability to behaviourally study these animals in natural settings restricts our overall understanding. It is likely that sharks use multiple sensory modalities simultaneously to achieve a desired outcome, and that the reliance on various sensory modalities changes between environmental conditions (e.g. water clarity, temperature), the focal item (e.g. different prey types) and age.

Vision

Shark eyes are often described as black, lifeless or vacant. However, behavioural experiments conducted on lemon sharks (*Negaprion brevirostris*) have shown that this species has a highly sensitive visual system, which could even be considered better than that of humans in certain characteristics (Gruber and Cohen 1978). The visual field of an animal is dictated by head morphology, eye position and mobility, pupil shape, lens and head movement, eye socket depth and the extent the eye protrudes from the head (McComb *et al.* 2009). The lateral placement of the eyes on the head, eye rotation and the common sinusoidal head movement pattern of sharks while swimming allows many species to have a nearly (or complete) 360° vertical and horizontal visual field; however, the degree of binocular overlap (which provides enhanced depth perception) may consequently be reduced.

Sharks have developed various adaptations to protect their eyes, suggesting the high importance and vulnerability of this sensory system. Some species (e.g. tiger sharks) possess an extra anatomical protective apparatus called the nictitating membrane. This 'third eyelid' extends from the lower nasal corner of the eye when the shark comes in near contact with another object (particularly during feeding) to cover the exposed eye. Sharks that lack this adaptation may instead use extraocular muscles to rotate their eyes cranially into the orbit for greater protection during periods of vulnerability (e.g. white sharks).

The photoreceptive component of the elasmobranch eye, the retina, is anatomically simple compared to that of many other aquatic and terrestrial vertebrates. While most species have both rod and cone photoreceptors, sharks have been found to have only a single class of cones (at most), suggesting that they are effectively colour-blind or are capable of only rudimentary colour discrimination (Hart *et al.* 2011). However, it is important not to confuse colour with brightness or contrast, which under certain circumstances can offer a similar discriminatory ability. The intricacies of the shark visual system have significant implications in shark attack mitigation.

Hearing

While sharks are not known to produce any sounds, hearing would provide both physical and biological information on their surroundings. Shark ears are only externally represented by two small holes on the head, called the endolymphatic ducts. The inner ear of sharks contains

A white shark (*Carcharodon carcharias*) with air bubbles escaping from one of its many sensory pores. Photo by Denice Askebrink.

a pair of labyrinths, each with three semicircular canals and four sensory maculae (Maisey 2001). Sharks are unable to detect sound pressure and instead rely on the particle motion of sound, including acceleration, velocity and displacement (Casper and Mann 2009). In field studies, sharks were most attracted to low-frequency, erratically pulsed tones that closely mimic the bioacoustic characteristics of injured or distressed fish (Casper 2011). Incidentally, this is also the acoustic footprint of boats. Sharks can localise acoustic stimuli from hundreds of metres; however, as the particle motion component of sound used by sharks is focused in the near-field (approximately one wavelength from the source) then quickly diminishes, sharks have been shown to only be responsive to ambient sounds (at biologically relevant volumes) from distances of less than 5 m (Casper 2011).

Hair cells within the semicircular canal have been shown to be sensitive to changes in ambient pressure, and thus could be used to sense both depth and atmospheric pressure (Fraser and Shelmerdine

2002). Blacktip sharks (*Carcharhinus limbatus*) have been observed to behaviourally respond to changes in atmospheric pressure related to tropical storms (Heupel *et al.* 2003).

Mechanosensation (distant touch)

The mechanosensory lateral line of sharks detects differential movement between the body and the surrounding water. It is used to detect dipole sources (e.g. prey) and uniform flow fields (currents). It also mediates rheotaxis (orientation relative to water current), predator avoidance, hydrodynamic imaging (avoidance of obstacles) and social behaviours, such as schooling and mating. There are two different types of canals – pored and non-pored. Pored canals are in contact with the surrounding water and extract information such as acceleration, and therefore water movement from predators, prey and conspecifics, and the animal's own location in relation to nearby objects. Non-pored canals are under the skin and may discern the velocity of skin movement through contact (Jordan 2008). The total number of lateral line pores and their exact distribution is highly species-specific, but in sharks they are typically found on the back and sides of the body and tail, behind the mouth, between the pectoral fins and behind the ear ducts (Hueter *et al.* 2004).

Electroreception

The elasmobranch electrosensory system is made up of subdermal groups of electroreceptive units called the ampullae of Lorenzini. These receptors are highly sensitive to low-frequency, weak polar electric fields created by living organisms. They are used to detect bioelectric fields of prey, predators and conspecifics, changes in temperature and inanimate electric fields, which may be useful in geomagnetic navigation (Hueter *et al.* 2004). Ampullary receptors are visible as small pores on the body. Within these receptors, sensory hairs and support cells are joined to provide a high-resistance electrical barrier, forming an electrical core conductor and functioning as voltage detectors (Tricas 2001).

Smell

The chemoreceptive ability of sharks relies on swimming pattern and speed, water movement, odour concentration, bilateral differences in

The ampullae of Lorenzini pores are clearly visible on the snout and around the eye of tiger sharks (*Galeocerdo cuvier*). Photo by Denice Askebrink.

signal intensity and arrival time, and the anatomy and morphology of the nasal canals and olfactory system. The smelling ability of sharks is so sensitive that they are able to compare odour concentration and arrival time between each nostril in order to orient themselves towards the source. This ability may be unique to cartilaginous fish (Hofmann and Northcutt 2012). External nares (nostrils) regulate water flow over elaborate and complex foliated olfactory tissue, designed to maximise sensory capability in a small area.

Taste

Sharks utilise taste buds to determine the palatability of items. While our knowledge of gustation in sharks is rudimentary, it has been shown that taste buds are found in the mouth, throat and gills. Taste buds are most concentrated in the front of the mouth, which suggests that they are utilised for taste assessment during the biting and manipulation of prey (Collin 2012). Interestingly, many sharks also possess oral denticles in the mouth and throat, which are used to enhance friction and grip on prey items during manipulation within the mouth and to protect the tissue from damaging hard-bodied prey items (Atkinson and Collin 2012).

Movement patterns

There is little information about sharks that would be more beneficial in mitigating the risk of shark–human interaction than having a thorough understanding of shark movement patterns. However, not surprisingly, their movement patterns are extremely complex, especially when multiple species and various regions are considered. Animals move for a variety of reasons; for sharks, these are thought to include foraging, mating, parturition and the use of nursery areas, and in response to certain abiotic environmental factors such as temperature and salinity. Changes in physiological requirements, energetic expenditure and risk of predation over ontogeny often result in changes in habitat use. Sex, health and personality also lead to differences in movement patterns, especially in relation to prey preference and risk-taking, creating significant variation among individuals within cohorts (Matich and Heithaus 2015). As such, exceptions have been found for pretty much every 'rule' in relation to shark movement. Overall, shark movement is most likely a result of temporal, circumstantial and inter- and intraspecific factors, making patterns only partly predictable.

In most animal populations, it is common for only a percentage of the population to migrate (partial migration), while other individuals remain resident in an area. This appears to hold true for (at least some species of) sharks (Holmes *et al.* 2014). Interestingly, and of particular note when considering motivation (for movement, and potentially for negative shark–human interaction due to increased energy requirements), genetic studies have shown that certain shark species are philopatric, returning to the same location to give birth over multiple years (Feldheim *et al.* 2004; Hueter *et al.* 2005).

While tagging and tracking studies can provide a wealth of knowledge on the movements of individual animals, these studies generally account for a very limited number of individuals or cover only short timeframes due to the significant resource requirements and degree of dysfunctionality of many tags. However, information on shark movement patterns is now being supplemented by various shark mitigation strategies, such as acoustic tagging and tracking, as in Western Australia and New South Wales (Australia). Although often presenting a localised picture, this information can be of great benefit in regional movement trends and in directing future regional shark attack mitigation strategies.

White sharks

White sharks have a circumglobal distribution, although they are primarily concentrated in South Africa, Australia/New Zealand and the north-eastern Pacific (Jorgensen *et al.* 2010). They are also being reported more frequently in the north Atlantic Ocean (http://www.ocearch.org/). Despite these localised activity areas, both sexes have been shown to be capable of trans-oceanic migrations.

Tagging and telemetry studies have shown that white sharks utilise a combination of four main movement patterns: rapid broad-scale trans-oceanic return migrations, long-distance directional swimming in coastal waters (generally at 5–100 m depth) at speeds of 2–3 km h^{-1}, smaller-scale patrolling and temporal residency/site fidelity (Bonfil *et al.* 2005; Bruce *et al.* 2006). The consistent aspect of white shark movement patterns suggest that they may follow established routes. However, there is also considerable plasticity in their movements, which is often attributed to the different foraging strategies and prey selection necessary to meet the requisite energy demands of immature, mature, male and female sharks (Bruce and Bradford 2015).

Prolonged residency and year-round visitation of male white sharks around major marine mammal rookeries has been well documented. The Neptune Islands support one of the largest colonies of New Zealnd fur seals (*Arctocephalus forsteri*) in Australia, and Seal Island off South Africa is a major Cape fur seal (*Arctocephalus pusillus pusillus*) rookery. These two locations also support white sharks year-round; however, individual sharks come and go, with typical residency periods averaging 11 days between migrations at the Neptune Islands. Sharks aggregate around seal rookeries in the austral autumn and winter when adult female and predator-naïve seal pups are most prevalent (Bruce and Bradford 2015). Males continue to visit marine mammal rookeries throughout the year, although their numbers at the Neptune Islands peak between December and February. Females are far less resident than males, segregating during the spring and summer. In South Africa, females typically migrate to in-shore waters of False Bay, Mossel Bay and KwaZulu-Natal, where migratory teleosts and elasmobranchs are more prevalent (Kock *et al.* 2013). They also migrate long-distance (>2000 km) along the continental shelf from the high-abundance areas in the Western Cape to KwaZulu-Natal, or further (Bonfil *et al.* 2005). Australian white shark migrations include broad-scale movements across their entire range, which extends up both the east and west coasts, around Tasmania and across the Tasman Sea to New Zealand. They likely follow migrating snapper (*Pagrus auratus*), tailor (*Pomatomus saltatrix*), mullet (*Mugil cephalus*) and salmon (*Arripis trutta*) (Bruce *et al.* 2006).

Sea surface temperatures have been suggested to influence white shark presence. However, as white sharks can tolerate a range of water temperatures by regulating their core body temperature, it is likely that this represents their preferred prey's physiological preference for warmer water, not necessarily the sharks' preference.

The transient and seasonal movement of sharks to in-shore waters shows a clear shark–human management concern. And indeed, although pinniped colonies may represent primary hunting grounds, they are not necessarily the most commonly used areas within aggregation sites. Aggregating white sharks at Mossel Bay were found to utilise core habitat areas of just over 1 km². Individual activity areas ranged from 6 km² to 56 km² and home ranges (defined as the day-to-day, spatial extent or outside boundary that the animal utilised) ranged from 3 km² to 22 km² (Jewell *et al.* 2013). Home range was negatively correlated with shark size, as larger animals utilised smaller home ranges and activity areas (Goldman and Anderson 1999).

Tiger sharks

Tiger shark movements are best described as confusing, arrhythmic and seemingly unpredictable. While tagging and tracking studies suggest some generalisations, the true generalist and opportunistic predatory nature of tiger sharks results in regularly observed behavioural anomalies. Tiger sharks are largely coastal; however, their distribution in the open ocean has been found to be considerably larger than previously thought (Domingo *et al.* 2016).

Tiger sharks are discretionary partial migrators that use both fixed intrinsic conditions (e.g. age and sex) and flexible extrinsic conditions (e.g. prey abundance and water temperature) to establish habitat use (Papastamatiou *et al.* 2013). Overall, tiger sharks are thought to have extremely wide home ranges (up to 634 944 km^2). Their movements and habitat usage are more regular in these areas and they return to these places on a regular basis between migrations (Heithaus *et al.* 2007b; Meyer *et al.* 2009a; Ferreira *et al.* 2015). Tiger shark migrations are generally rapid, directed and predictable long-distance movements of 100–1000 km along coastal shelves, but can also be far greater across open oceans (Heithaus *et al.* 2007b; Werry *et al.* 2014). They are known to follow coastal currents (Hazin *et al.* 2008). During migrations, tiger sharks can maintain average speeds of up to 3.5 km h^{-1} for over 8000 km.

The movement patterns displayed by tiger sharks suggest that they may utilise cognitive maps to navigate among distant foraging areas, with variation in individual spatial behaviour attributed to experiential exploration and learning (Meyer *et al.* 2010). Both orientation (the moment-to-moment alignment of the body) and navigation (directed movement towards a goal) involve neural processing of sensory inputs for directional and distance determination and both are dependent on the formation, storage and retrieval of spatial memories, allowing for the return to fixed environmental points (Guttridge *et al.* 2009).

Justification for the diversity in tiger shark movement patterns revolves mostly around the element of stealth and surprise that this species largely relies on for predation. Once potential prey in a given area become wary of a shark's presence, the element of surprise is lost and the shark moves on. Differing movement patterns may also reflect variation in foraging strategies necessary for exploiting different prey types, or to take advantage of seasonally abundant naïve prey.

Significant intraspecific variation in tiger shark movement is also clear. Juveniles in Hawaii displayed wider ranges and were detected less frequently and for shorter periods of time compared to mature females. This was attributed to juveniles' greater need to avoid predators and explore new areas to establish individual home ranges (Meyer *et al.* 2009a). In contrast, subadults and an adult male shark were residential in the Chesterfield Islands in the Coral Sea, whereas adult females underwent long-range migrations to the Great Barrier Reef (Australia), potentially connecting spatially segregated populations (Werry *et al.* 2014).

Of particular note, Meyer *et al.* (2009a) found that despite the normal pattern of tiger sharks staying in a given area for only a matter of minutes, some sharks were observed to display unusual periods of residency around a particular harbour where there was frequent disposal of fish carcasses. The potential for such behavioural shifts, from wide-ranging movement patterns aimed at targeting risk-averse prey to residential behaviour facilitated by scavenging near shore, could have serious implications if such locations overlap with human use areas.

Bull sharks

Bull sharks are another cosmopolitan species, whose movement patterns are of great interest to shark attack mitigation strategies. However, similar to those of white and tiger sharks, their movements tend to be complex and variable. In relation to shark–human activity, bull sharks' ability to easily cross salinity barriers extends their overlapping range with many human use areas, such as canals, estuaries and rivers. Short-term tracking studies of these sharks in both the Atlantic and Pacific oceans have shown that bull sharks display some site fidelity to certain coastal and continental shelf regions but also undertake frequent substantial coastal migrations (Brunnschweiler *et al.* 2010; Daly *et al.* 2014). Sharks tagged along the south-east coast of the US have travelled west to Texas and north to South Carolina (Carlson *et al.* 2010). Further support for bull shark migration stems from genetic analyses, which indicates gene flow between more remote locations (Lea *et al.* 2015), and from simple seasonal presence/absence studies using fisheries or shark attack mitigation data. Bull sharks have been shown to be capable of average sustained speeds of 0.7 km h^{-1} and maximum speeds of 2.5 km h^{-1} (Daly *et al.* 2014). Bull sharks in Mozambique exhibited similar patterns of temporal and spatial residency interspersed with seasonal, largely coastal migrations of 450–3760 km at an average rate of 18 km/day. Male bull sharks in southern Africa were found to be slightly more resident than females, which may reflect the theory that pregnant females often take advantage of river or estuarine habitats for pupping, even travelling distances of 4000 km round-trip across open oceans to do so (Lea *et al.* 2015).

Bull shark migration and residency patterns are often attributed to water temperature; however, the diversity in findings suggests that other factors, such as reproductive activity and foraging, are also relevant. For example, residency has been shown to correlate with large aggregations of preferred prey, such as trevally in southern Africa (Daly *et al.* 2014) and kingfish and tuna in Sydney Harbour (Smoothey *et al.* 2016).

Blacktip sharks

Blacktip sharks (*Carcharhinus limbatus*) contribute to a large proportion of the bites in Florida. Thus, having an understanding of their movement patterns clearly benefits mitigation in that region. Blacktip sharks are most prevalent in Florida in the winter (January–March), when the Gulf Stream current converges with the narrow coastal shelf near Palm Beach County. During these times, shark abundance can exceed 12 000 individuals in the 76 km^2 area, equating to 803 sharks/ km^2 (Kajiura and Tellman 2016). Here, shark abundance is inversely correlated with water temperature, with peak numbers encountered only when temperatures drop below 25°C. Shark abundance is correlated spatially and temporally with spawning aggregations of baitfish. Over the course of the year, this population migrates as far north as North Carolina, or potentially even Delaware Bay, before returning southward in autumn.

Conclusion

Sharks are a natural and integral part of the aquatic ecosystem and are important to humans in many ways. They are impeccable predators, with cognitive and sensory capabilities on par with or superseding those of mammals, birds, reptiles and amphibians. Sharks have the biological and physiological capacity to locate, capture and kill some of the largest, most mobile, defensive and intelligent animals in their environment; an environment where humans are awkward and inefficient. Therefore, given the amount of time humans spend in the water, if we really were on the menu, we would know about it. We are not an unpredictable species in terms of our coastal waterway usage, yet shark residency and migration behaviour around high human use areas is strikingly in contrast to that of marine mammal rookeries, for example. The difference in white shark presence between Adelaide beaches and the Neptune Islands in Australia, or San Francisco and the Farallon Islands could not demonstrate this discrepancy more clearly. All these places have high mammal populations but sharks characteristically frequent the islands with marine mammal populations and show much greater residency periods at those locations, whereas they are far more transient in areas with reliably high terrestrial mammal (i.e. human) populations.

3

Shark attacks, deconstructed

Fact: occasionally, sharks bite people. Fact: sometimes shark bites are fatal. However, the fact also remains that the likelihood of being bitten by a shark, without provocation, is incomprehensibly low in most situations and fatal bites are even more rare, and are becoming increasingly so.

Rhetoric, myths and reality

For some reason, to blow things completely out of perspective, just add water. Ask any fisherman how big 'the one that got away' was, and the size is almost guaranteed to be at least a third larger than what it actually was. Granted, water does optically magnify objects. This aquatic trend also seems to hold true for our perception of shark attacks – the perception, risk and fear of such events always seem to be blown far out of proportion.

There are many clichéd statistics around a human's probability of being attacked by a shark. The ones that come to my mind are that you are more likely to be struck by lightning than be attacked by a shark, more likely to be killed by a vending machine and more likely to die from a bee sting. A quick Google search of 'how likely am I to be attacked by a shark?' returns around 2 170 000 results. Google's key finding quotes the International Shark Attack File and states, 'Your chances of being attacked by a shark are just one in 11.5 million … On average, there are ~65 shark attacks worldwide each year; a handful are fatal. You are more likely to be killed by a dog, snake or in a car collision with a deer.' But after that helpful, researched and direct answer, the results spiral into a long list of articles and websites that predominantly focus on a random but convenient number of often 'shocking' things more likely to kill you than a shark. A subsection of the information that I found, all relating to things more threatening to human life than sharks, included falling coconuts (150 deaths per year), secondary school and college football (12 per year), champagne corks (24 per year), tripping and falling at home (6000 per year), choking on food (3000 per year), lightning (24 000 per year), jellyfish (40 per year), and attempting to take selfies (40 per year). Just for reference – in 2015, six people were killed by sharks.

Further sites relating to annual US statistics, alone, suggest that 100 people die a year from scalding-hot tap water, 15 from falling icicles, six from spider bites, 450 from falling out of bed, 355 from falling off a ladder, 5000 from consuming contaminated raw meat, 20 from horses and 20 from cows, 100 from bee stings and 2000 as a result of incidents with traffic lights. There was one shark attack fatality in the US in 2015.

Through my Google search, I also learned that a person has a 0.0000000004891% chance of being attacked by a shark while being struck by lightning (Hart 2015). I'm yet to figure out why this statistic is significant, but if it has happened to someone, at some point during human history, then I greatly feel for that person. I fear that they are (or were) probably the unluckiest person on the face of the planet.

While entertaining, the problem is that these 'facts' may or may not be correct. In the same set of searches, the statistic that hippopotamuses kill around 2900 people a year also surfaced. In fact, I stopped counting at 25 individual web articles that quoted that statistic. Yet, despite extensive searches through a variety of more scholarly outputs, I was unable to dig up information to equate to anywhere near that number. However, interestingly, and as an aside on my newly gained knowledge on hippopotamuses, the IUCN Red List of Threatened Species estimates the current population of common hippopotamuses (*Hippopotamus amphibius*) at 125 000–148 000 individuals (Lewison and Oliver 2008). Taking the average of this range, and the commonly quoted internet statistics, this would mean that one in 47 hippos is responsible for a fatal attack on a human every year, unless of course dwarf and pygmy hippopotamuses, which were not included in the earlier population figure, account for a significant proportion of fatal attacks on humans.

My guess is that this repeatedly quoted 'statistic' is highly inaccurate. And I imagine that many of the statistics related to shark attack are equally false, misinformed or misleading.

Many of the most prevalent media articles returned from Google searches do not provide references of any sort detailing the source of their information. When references are cited, they often link back only to other media reports. While it should not be immediately or absolutely assumed that these statistics are false, they should also not be automatically accepted as fact. More than likely, they are just part of a cycle of regurgitated rhetoric from time-poor journalists.

When more trustworthy sources are consulted, which provide research-based statistics, it becomes clear that there are truly many things more likely to kill you than a shark, including the traditionally cited bees, wasps, snakes and lightning. Trauma is more likely to result during travel to or from the beach, from spinal damage during water-based recreational activities, jellyfish stings and stingray barbs, sunburn and seashell lacerations (ISAF 2017).

Even more relevant, and alarming, are the statistics reported by the US Center for Disease Control and Prevention. These statistics show that annually, over recent years in the US alone, 614 348 people died of heart disease, 591 699 people died of cancer, 128 978 people died of cardiovascular disease or stroke, 33 804 people died from motor vehicle accidents and 16 121 people were murdered by another human. We should find this far more confronting than the yearly average of less than one shark attack fatality over the same period. Yet we don't. For some reason, we have come to terms with terminal diseases and murder but not with shark attacks.

In more global terms, but of similar relevance, is the fact that in 2000 the estimated global number of (recorded) mortalities from drowning was a staggering 409 272 (WHO 2016). This figure is second only in global unintentional injury deaths to road traffic injuries. The global number of shark attack mortalities in 2000 was 11 (ISAF 2017).

To finish on a statistically positive note – you are more likely to win an Oscar, a Nobel Prize or a lottery jackpot of US$1 million or more than to be fatally attacked by a shark.

Sharks compared to their terrestrial counterparts

Fatalities attributed to wild animals are in no way unique to sharks, and indeed many civilisations (coastal or inland) conflict with at least one species of potentially dangerous animal. Official records from Mozambique detail 265 mortalities attributed to wildlife between July 2006 and September 2008 (27 months) (Dunham *et al.* 2010). Nile crocodiles (*Crocodylus niloticus*), lions (*Panthera leo*), African bush elephants (*Loxodonta africana*) and common hippopotamuses were most implicated, but buffalo (*Syncerus caffer*), spotted hyenas (*Crocuta crocuta*) and leopards (*Panthera pardus*) were occasionally involved. The annual fatality rates of these species were 59.6 for crocodiles, 13.7 for elephants, 10.7 for lions, 5.3 for hippopotamus (not quite a significant contribution towards 2900, despite Mozambique being a current conservation stronghold for the species) and 0.4 for buffalo. In comparison, no shark attacks occurred off Mozambique during the same period, fatal or otherwise, and the corresponding annual average of shark-related fatalities (for all 509 shark species) worldwide was just 4.9 (Shark Research Institute 2016).

An examination of the US National Center for Health Statistics over the 11-year period of 1991–2001 found that 1943 people were fatally injured in the US as a direct result of interaction with animals (excluding traffic accidents caused by animals, zoonotic disease transfer and riding accidents), equating to an average of 177 per year (Langley 2005). The most implicated species were hornets, bees and wasps (48.5 per year), dogs (18.9 per year), spiders (6.0 per year) and snakes (5.2 per year).

Further (referenced) statistics reported in a blog by Bill Gates (Gates 2014) on the world's most dangerous animals showed that, annually, parasites and vector-borne carriers account for far more fatalities. These include tapeworms (2000), ascaris roundworms (2500), freshwater snails (schistosomiasis, 10 000), assassin bugs (Chagas disease, 10 000), tsetse flies (sleeping sickness, 10 000), dogs (rabies, 25 000) and, most notably, mosquitoes (malaria, dengue fever, yellow fever, encephalitis, 725 000). Sadly, the second deadliest animal to humans, responsible for 475 000 fatalities annually, however, is humans.

Shark attack files

Our ability to analyse and learn from shark attack incidents is possible due to a few expertly maintained shark attack files. The two major files are the Global Shark Attack File and the International Shark Attack File. They operate on the premise that through gaining a better understanding of these events, we might be able to lessen their future likelihood. The Global Shark Attack File (GSAF) is run through the Shark Research Institute (Princeton, US), while the International Shark Attack File (ISAF) is owned by the Smithsonian Institution, housed at the Florida Museum of Natural History and under the trustee and management of the American Elasmobranch Society (Gainesville, US). Both files invite submissions of suspected shark attack cases, and investigate these and independently researched incidents, utilising authorised investigators around the world. Each file has a compendium of around 6000 cases that date back to the mid-1500s (and a few cases from even earlier, in the GSAF). Unsurprisingly, there are some discrepancies between the two files, with cases in one not always mirrored in the other. It is readily acknowledged that not all shark attacks are reported, that information on shark attack statistics from developing countries is particularly lacking (and thus underrepresented), and that some locations make concerted efforts to keep information on attacks quiet to minimise bad publicity (ISAF 2017). The ISAF appears to be more conservative and tends to not include questionable, unproven or inconclusive incidents, whereas the GSAF is more inclusive of suspected cases, aiming to capture all possible incidents. It contains 4250 confirmed records of unprovoked shark attacks between 1543 and 2015. The Australian Shark Attack File (ASAF), based at Taronga Zoo (Sydney), has been in operation since the mid-1980s with a similar aim, but it includes shark attacks in Australian waters only (ASAF 2016). While the files can't be expected to be exhaustive, they provide impressive and highly useful banks of information to support shark attack research, enable the examination of long-term trends and, not insignificantly, provide well-informed contacts, factual data and reports to the media.

The statistics on shark attacks in this book are extracted from the information freely available from these shark attack files.

Shark attack, literally

Through my employment, I have caught sharks, injected sharks with medicines and taken intravenous blood samples, and I've hand-fed, cornered and wrestled sharks. As a result, I have been bitten by sharks (although, luckily, each bite has been very minor). Yet you will not find my name in any of the shark attack files. This has to do with a highly relevant aspect of shark attack terminology, which accounts for the backstory and reason for the interaction. Every time I was bitten, it was because I was forcing interaction with the shark, preventing its natural behaviour and/or was too close to its food. I was purposefully interfering with the shark, causing stress and sometimes pain. Therefore, the response of the shark – whether it was to swim away or to bite me – was my fault, and my responsibility, not the shark's. The key terminology here is 'provoked'. While provoked attacks may still be reported in shark attack files and are sometimes investigated, they are always clearly identified as such and are excluded from the far more relevant and meaningful 'unprovoked' shark attack statistics and analyses. In addition to the aforementioned modes of provocation, others that may result in shark attacks include spearing or otherwise fishing for sharks, especially when trying to remove a hook, stepping on, poking or pulling a shark's tail, or attempting to ride a shark.

Other key discriminating terms relating to shark attack statistics include 'boat attack', which involves a shark making contact with a part of a motorised or non-motorised boat or vessel; 'war' or 'air–sea disaster', which involves humans being in the water because of an act of war (e.g. an aircraft being shot down, or a warship being sunk) or a recreational or commercial sunken ship or downed aircraft; 'scavenge', which reflects post-mortem bite or consumption; and 'insufficient evidence' or 'questionable cases', which lack sufficient data to determine the presence or accountability of a shark in the event. These cases are also excluded from 'unprovoked' shark attack statistics.

The nature of shark attacks

Shark attacks are not uniform events, and the ISAF utilises three classifications to describe unprovoked shark attacks. The first, and by far the most common (constituting up to 80% of attacks), is the 'hit and run' attack. These attacks are characterised by a shark making a single act of contact, usually a bite or slash, and not returning. The injuries to the victim from this type of attack are usually minor, often involving lacerations to the leg below the knee, and are seldom life-threatening. These attacks most often occur in the surf zone on swimmers and surfers, and often in water affording poor visibility or in harsh environmental conditions (e.g. breaking surf, strong current or turbulent water). As these attacks happen so quickly, the victim rarely sees the animal involved. Sharks must make rapid decisions during predatory attacks in order to capture their typical prey and, under suboptimal conditions, sensory signals emitted by humans, for instance through splashing and erratic movement, light reflections off jewellery or contrasting body colouration (e.g. relatively tanned legs but pale soles of the feet), could easily be misinterpreted as prey cues. The single bite may provide the shark with enough information to realise that it has not encountered an optimal prey item, so it does not return. Hit and run attacks may also be unrelated to predation; instead, they may be used as a social assertion of dominance if the shark feels threatened in its environment. Other theories speculate that this type of attack represents an immature predator strategy of juvenile sharks, even to the point of being referred to as petulance (Baldridge and Williams 1969). The main assailants in this type of attack are smaller reef sharks, with blacktip sharks, and possibly spinner (*Carcharhinus brevipinna*) and blacknose sharks (*Carcharhinus acronotus*), the most notable (ISAF 2017).

The second and third categories of shark attacks are 'bump and bite' and 'sneak' attacks. These are the attacks that lend themselves to the scripts of horror films and nightmares, and are the reason that so

I consider myself a part of the *Jaws* generation. I didn't see the movie until 1981, but it was such a big thing in my life and it scared the hell out of me, so I remember it really well. I grew up on Long Island, New York, surrounded by water, and I was deathly afraid of the water after that. It didn't matter if it was pool water, or the ocean; if anything broke the surface, I was convinced that it was a shark coming to attack me. I watched the movie again when I was 15 or 16, and this time I loved it. It turned into a fascination and it made me want to be a marine biologist and study sharks. So I ended up going to Southampton College on Long Island because they had a program called Seamester, where you took classes over a nine-week period while aboard a boat that sailed from the Caribbean to New York. I finally got to take part in the program, and six weeks in, we got to the island of Rum Cay, where we were going to look at swamp ecology. We took a small one- or two-person kayak to the island with us, which at the time I thought was kind of stupid. After all, there were 20-something kids in the water, plus a couple of professors and boat crew. We were walking across a small salt water lake as a group and I was on the far right. The water was maybe mid-thigh. I heard the water rushing, and from ~6 or 7 feet away, I saw the shape of a shark coming at me. And then I just felt an impact that drove my leg back. Even though I had seen the shark, I still didn't comprehend what had happened, and assumed I had walked into a rocky outcropping or stepped on a stick. My mind had not caught up with what I had seen. I looked down, but we had kicked up a lot of silt, so I couldn't see anything. But then someone commented that there was a lot of blood in the water, and that's when I reached down, and where I knew I should have been feeling solid flesh on my leg, I wasn't. I could feel the warmth of my blood flowing over my fingers. One of the guys on the trip was a former Eagle Scout, and the first thing he did was call for everyone's t-shirts, which he wrapped around my leg. They brought the kayak over, which turned out to be a good thing, and they carried me on the kayak back to the mainland. There are only ~75 people that live on the island year-round, but luckily they found a guy who had the only plane and the only car on the island. I was loaded into the back of his jeep and taken to his house. By that time, they had radioed for the ship's medical officer. When he peeled the shirts back, my two friends saw the injury and they both just burst into tears, so I knew it wasn't good. I was flown over to Long Island, where the nearest doctor was. The injury wasn't necessarily painful, but definitely uncomfortable. It felt like a blunt trauma, and I could feel the pressure of the blood coming out. When I got to the doctor, they gave me painkillers and told me I needed to go to the hospital at Nassau to get stitched up. It was only at that time, under heavy sedation, that I looked at my leg, and so I only have a very hazy recollection of what it looked like. We made it to the hospital at Nassau around five hours after I was bitten, and then I underwent a 2.5–3 hour operation where they stitched up both the inside and outside of my leg. I later found out a lot of what they had done was for irrigation, just to really make sure the injury was completely cleaned out. I don't know if the surgeon had had previous experience with fish or shark bites before, but they were concerned with bacteria, so they wanted to make sure it was cleaned out really well before they stitched it up. I flew back to New York four days later.

On my next medical assessment, I was told there was good news and bad news. The good news was that I didn't need physical therapy; the bad news was that it was because physical therapy wouldn't do anything. All of the damage was to the outside of my right leg. From the middle of the shin back, I don't have any feeling because all of the muscles and tendons were cut. I had a custom orthotic brace made, which I was told I'd need to walk. But after around a year of wearing the brace full-time, I started taking it off little by little, and teaching myself to walk without it. So instead of heel to toe, I started walking by lifting from my hip and knee, and you can't really even tell I'm doing it anymore.

I think it was probably a juvenile lemon shark, about 3 feet long. They found out later on that the people who live on that island don't go into that body of water because it's a known lemon shark breeding ground. I imagine that with 20-something kids going through a body of water that doesn't normally get much action, we probably spooked the shark, and I must have inadvertently trapped the shark between myself and a rocky outcropping. It probably felt cornered, so it bit me in an effort to escape.

Sharks still don't bother me, but they are in the back of my mind. I swim in the bay here occasionally, but every once in a while, I get a tingling feeling, and I'll get out of the water. Through my work in the field of marine biology, I talk to kids every year about shark attacks during *Shark Week*, and it's interesting what these kids learn and know. A lot of it is from the Discovery Channel, and some of its good but some of it isn't. It's often like I have to deprogram the kids from this information.

Scott Curatolo-Wagemann

many people fear sharks. Bump and bite attacks are characterised by the shark initially circling and bumping the victim before an actual attack, thus spending a much greater amount of time in the vicinity (and mind) of the person. It is thought that, during these situations, the shark may be assessing the potential danger of a prey item or even attempting to inflict injury or incapacitation before carrying out a more focused attack (McCosker 1985; Baldridge 1988). Sneak attacks, on the other hand, are as the name implies; the attack happens rapidly, without warning, and literally out of the blue. In both bump and bite and sneak attacks, the shark commonly carries out repeated bites and more elaborate injuries are often sustained. While these types of attacks are less common, the most typical victims in these attacks are divers or people swimming further off-shore, and the resulting injuries are often severe and potentially fatal. Bump and bite and sneak attacks most often take place in deeper water, but may occur in near-shore waters. White (*Carcharodon carcharias*), tiger (*Galeocerdo cuvier*) and bull sharks (*Carcharhinus leucas*) are most notorious for these sorts of attacks; however, other species are likely to also employ this behaviour, such as great hammerhead (*Sphyrna mokarran*), shortfin mako (*Isurus oxyrinchus*), oceanic whitetip (*Carcharhinus longimanus*), Galapagos (*Carcharhinus galapagensis*) and Caribbean reef sharks (*Carcharhinus perezi*) (ISAF 2017). These attacks likely represent more considered actions by the shark, and are targeted feeding or antagonistic behaviours.

Shark attack statistics

It is well acknowledged and reported that shark bites are increasing in prevalence globally; a trend that is due to the amalgamation of a range of different factors. Arguably, however, the most relevant and highly causative factor leading to increased shark–human interaction is the global increase in human recreational water usage. There has been an upsurge in human population, and humans are drawn to the sea for residence, tourism and recreation. However, population and water usage increases alone do not entirely explain the spatial and temporal surges in shark bite statistics. It is more likely that a combination of factors disrupts the natural balance of a region over a certain period, and increases the probability of interaction. Other factors likely to be involved include habitat destruction/modification, water quality, climate change, anomalous weather patterns and the distribution/ abundance of prey (Chapman and McPhee 2016).

Although there is a clear trend showing that shark bite incidence is increasing on a decadal scale, it is not the case that incidence levels rise on an annual basis. When grouped by decade, drastic differences in average incidence rates are seen: 49.9 per year in 1990–1999, 66.1 per year in 2000–2009, and 81.9 per year in 2010–2016. However, while 2015 had the highest recorded incidence with an uncharacteristically numerous 98 attacks, the next highest year on record was 2000, with

87 incidents. These years were separated by clear troughs of just 55 in 2008 and 56 in 2003.

The US consistently has the greatest number of shark attack incidents per year, often doubling those of the rest of the world combined. Within the US, Florida records the most shark bites annually, by far. In fact, shark bites in Florida waters generally comprise 30–40% of worldwide annual incidence. Within Florida, a disproportionate number of shark bites occur in Volusia and Brevard counties, whereas Flagler County, which borders Volusia County to the north, records very few shark bites. The spatial discrepancy in shark bite prevalence between counties is likely due to regional variation in water quality, natural and man-made environmental conditions, water movement patterns and beach visitation. The beaches of Flagler County are less popular, and in some areas are covered in exposed rock. In contrast, Volusia County beaches are far more populated. Man-made rock jetties and channels have altered natural water movement, creating conditions that are highly favourable for surfers. Furthermore, heavy rainfall and strong water movement in the area often cause decreased water clarity.

Historical shark attack statistics

Although seemingly a recent phenomenon, accounts of shark attacks reach as far back as recorded communication goes. An Aboriginal Australian rock carving at Bondi thought to be up to 2000 years old depicts a male figure, described as an iguana or lizard man, being attacked by an 8 m shark (Meadows 1999). Numerous references to shark attacks appear in ancient Greek and Roman accounts and 19th-century US texts, detailing events from the mid-1700s. Case 0000.0336 in the Global Shark Attack File, dated 336 BCE, details an attack off a protected bay in Greece: 'While a candidate for initiation was washing a young pig in the haven of Cantharus, a shark seized him, bit off his lower parts up to the belly, and devoured them.'

Although shark attack incidence is increasing, the rate of fatality as a result of shark bite is decreasing, with just six worldwide in 2015 (6.1% of all unprovoked bites) and only four in 2016 (4.9%). The 2016

I have been with Volusia County Beach Safety for almost 20 years, and I've treated ~15 shark bites during this time. Volusia County is well known for its shark activity, but my experience is that these are typically minor bite and release cases of mistaken identity, where the shark bites and immediately swims off. We only ever refer to them as bites, they're not attacks. Most are surfers, who are aware of the risk, and say they will not be deterred from surfing or going into the ocean, and most are treated on scene and not transported.

Following a bite, the typical victim will get out of the water on their own and flag down a lifeguard, or sometimes just leave. It is not as traumatic as what you see on TV. The shark bite victims I have treated are like any other minor medical call, involving basic life support. Some have had more serious lacerations, but none life-threatening. We have had to deal with far more serious medical and law enforcement calls for other reasons. Still, everyone reacts differently when bitten. Some are shocked and can't believe what happened and others are excited that they have a good story to tell. I usually have to deal with media following any shark bites due to the sensationalism from it, but I don't think that this is impeding, as I can usually get all the facts out and stop any false rhetoric. We are also asked to take photos of the wounds and fill out a shark bite survey for the International Shark Attack File, which includes a long list of questions relating to the event.

I am an avid recreational water user and am usually in the ocean five days a week in the summer time. Nothing has changed since treating shark bite victims and I actually have less of a fear of sharks now than what I did before working here. However, it is still always a thought in your mind. All of my co-workers swim/surf in the ocean most days without a major fear of sharks, and none of them have been affected by witnessing or attending to bites, either.

I think that most people who come here are aware of the risk, especially the locals. We have extremely large crowds all summer long. On occasion I get the question, 'Are there sharks in the water?', but the potential presence of sharks does not seem to deter water activity. I don't feel that people avoid our beaches because of the history of shark bites here, and we have had record crowds on holiday weekends and spring break over the past couple of years. From what I have seen, there has been very little change in shark and/or bait fish (or other prey) activity or environmental change during my time in the job. We do have some years with more bites than others, but nothing that I can point out that has been too significant.

We don't currently have any restrictions or bans as a result of the potential for shark bites. The inlet is where most of our shark bites occur, and we fly the purple flag there and have signage that states 'Hazardous marine life have been spotted in the area'. But the inlet is one of the best surfing destinations on the east coast of the United States, and surfers flock to this particular area. The hazard is compounded by the abundance of bait fish often found near the jetty. I can't speak for other parts of the country and world, but I don't think that any sort of shark attack mitigation measures are necessary here at this point. And again, not speaking for other areas, but here in Volusia County, we do not have any areas that are unpatrolled. All of our beaches are patrolled with lifeguards and emergency medical technicians with a response time of usually under two minutes. However, if you were to happen upon someone who had been bitten by a shark at a location that did not have such services, then the first thing you should do would be to immediately call for medical assistance.

Senior Captain Tamra Marris
Volusia County Beach Safety Ocean Rescue

fatality rate was *less than half* its inclusive 10-year (2007–2016) average of 11.5%. Based on information from the Global Shark Attack File, and summarised by Ricci *et al.* (2016), since the 1900s, 24% of shark attacks have been fatal and 76% were non-fatal; victims averaged 26.1 years of age; 80% were male, 10% were female and 10% were unknown; and the percentage of fatalities has decreased from 60%. The majority of attacks resulted in a single bite event (56%), 7% resulted in two bite events (the shark bit, left, then came back and bit again) and 0.5% involved three bite events. Anatomically, legs were the most common place for injuries (42%), followed by arms (18%), the torso (7%) and the head/neck (2%). These numbers are not surprising, in that legs are the most common body part in the water. They are what is below the surface when people are wading in the surf, and they dangle off surfboards while surfers sit on their boards waiting for waves. Legs also trail behind swimmers and divers, allowing them to be targeted without the shark being seen before an attack. Arms, as the other extremity, comprise the next largest group. These may also dangle in the water when surfers lie on their boards. Further, hands and arms are the most prevalent human 'weapon' used against sharks. Limb loss occurred in 7% of cases. Interestingly, the significantly extracted characteristics that increased the probability of fatality included swimming, three or more bite sites, limb loss, or attack by a tiger shark.

Shark attack, by location
Shark attacks have been recorded from six continents (all but Antarctica), 97 countries, 23 territories and independent islands, three

oceans and multiple seas. However, of these 120 countries and territories, 37 (a notable 30.8%) have recorded only one incident. While the US consistently records the greatest number of shark attacks per year, the percentage of fatal bites is very low. Australia and South Africa record the next highest numbers of bites annually (and the highest numbers of shark bite fatalities), followed by Brazil. New Zealand, Papua New Guinea, Réunion Island, Mexico, the Bahama Islands and Iran round out the top 10 localities for historical shark bite incidence.

There are currently 194 countries in the World Atlas, half have had confirmed cases of shark attacks and 64.4% of the 149 countries that have coastal borders have recorded shark attacks. However, while the latter statement provides a seemingly more relevant reference, these statistics still need to be interpreted with care, as coastal marine environments do not represent an exhaustive limit for shark habitat. Bull sharks are very well known to readily traverse salinity boundaries, and have been observed as far as 4000 km up the Amazon River in South America. They are also found, and known to breed, in Lake Nicaragua, a freshwater lake in Central America (Compagno 1984; Handwerk 2005). Bull sharks have been recorded in the Mississippi River as far north as Illinois, in the Ganges River in India, and in several other major global river systems (Compagno 1984). River sharks within the genus *Glyphis* also occupy freshwater habitats. While previously thought to be confined to freshwater, and thus contained to select regions, more recent studies have indicated that these species are euryhaline, with juveniles utilising river environments while adults are oceanic (Li *et al.* 2015). Thus, this generally cryptic species could be more widely distributed in freshwater river systems around the globe than previously thought. Not surprisingly, shark attack statistics mirror shark habitat usage; there is a confirmed case of a fatal unprovoked shark attack in Paraguay, which is firmly landlocked by Brazil, Bolivia and Argentina, and more than 1000 km away from the Atlantic Ocean via the Paraná River.

Although the topic is complex, diverse and includes a large degree of randomness, some trends in regional shark attack statistics can be identified. Chapman and McPhee (2016) analysed trends in shark attack statistics from identified global 'hotspots' (the US, South Africa,

Australia, Brazil, Réunion Island and the Bahamas) to extract potential drivers of increased shark–human interaction at those locations. The high rate of shark attack in the US (52% of globally recorded unprovoked shark bites between 1982 and 2013) compared to the rest of the world is certainly notable, and the incidence rate is rising at ~1.07 bites per year. The authors attributed the high prevalence and increase in incidence rate to an overlap in regional natural and man-made conditions that are preferable to both humans and certain species of sharks (predominantly juvenile Carcharhinid species and other relatively small reef sharks), the growing human population and beach usage, and extremely efficient communication.

Australia regularly records the second-highest number of unprovoked shark bites annually. Australia is a large country with a very long coastline, and its natural ecology and diverse aquatic environments support a wide range of shark species. Indeed, Australia represents the largest Chondrichthyan hotspot, with the greatest amount of species diversity (Weigmann 2016), including around 180 species of sharks. The locations where people are bitten and the severity of shark bites in Australia mirrors this spatial range and species diversity, as well as that of the human population. Recent rises in shark bite incidence in Australia have generally been attributed to increased human population and human beach usage, increased interest in water-based recreational activities, and increased visitation to previously isolated and underutilised beaches (West 2011). However, additional factors have also been suggested. There was decreased shark attack incidence during heightened cyclone activity, whereas spikes in incidence were observed during periods of extreme and protracted heatwaves, suggesting variation in shark attack prevalence around changing climactic activity and anomalous weather events. Recent surges in the numbers of previously overexploited but highly preferred prey items for large sharks, such as humpback whales (*Megaptera novaeangliae*) and New Zealand fur seals (*Arctocephalus forsteri*), may be drawing more sharks to the region (Chapman and McPhee 2016).

Although identified as a global hotspot for shark attack due to the average of 4.4 incidents per year, the average incidence rate of shark attack in South Africa has remained relatively unchanged over the last few decades. The relatively high number of fatalities in South Africa (compared to the US) is due to the difference in shark species that frequent and aggregate in South African coastal areas; namely, white sharks. White sharks are known to reliably congregate around the south Western Cape; unsurprisingly, this is mirrored in the higher proportion of bites in that region (39% in the Western Cape Province and 43% in the Eastern Cape Province, compared with 18% in KwaZulu-Natal). White shark activity also increases in the in-shore waters of Cape Town during the spring and summer months, when food sources (e.g. whales) are also present (Nel and Peschak 2006). Analysis of the results from South Africa highlighted another key variance in shark attack statistics, when the number of unprovoked bites jumped to 16 in 1998. Back-studies identified multiple anomalous weather patterns throughout the Indian Ocean at this time, attributed to the El Niño phenomenon. The result was cooler than normal sea surface temperatures in the eastern and equatorial Indian Ocean, while the western Indian Ocean measured warmer temperatures (1–2°C above long-term averages). Additionally, abnormally low rainfall was recorded in South Africa, whereas other locations had higher than average rainfall (Murtugudde *et al.* 2000). While these environmental changes cannot be definitively linked to the increase in shark attack incidence rate during and following the climactic event, changes to rainfall, sea surface temperature and disruptions to current patterns can lead to significant alterations to upwellings of nutrient-rich water, variation in biological productivity and prey availability over the entire food chain and, consequently, species usage of particular environments and currents. Subsequent El Niño patterns have also been cited as a factor in increased shark attack incidence in other parts of the world.

While Brazil and Réunion Island do not record anywhere near the same number of bites annually as the US, South Africa or Australia, recent changes in shark bite statistics are particularly notable in those places. Following a long history of no (or exceptionally few) shark bites in Brazilian waters, bites suddenly became a regular occurrence beginning in 1992 and a peak of 10 unprovoked bites was recorded in 1994 (Chapman and McPhee 2016). Similar to the regionality of shark attack incidence in the US, bites in Brazil occur almost entirely along a very small (20 km) span of coastline in Recife (Pernambuco state). This clear delineation from no-incidence to high-incidence presented an extremely valuable opportunity to examine what changed, and the best chance of identifying a definitive causative factor (or factors). Interestingly, virtually all the identified changes can be directly attributed to human action through the development of a commercial port in Suape. To facilitate the new port infrastructure, the local environment was dramatically altered – coral reefs were removed, deepwater channels were excavated, mangrove habitat was destroyed, river mouths were altered or closed and estuaries were diverted. These changes could easily alter the typical composition of permanent inhabitants and the normal behaviours and patterns of transient residents in that environment. Of particular regional interest was the local bull shark population, which had previously utilised the river mouths that were altered or closed off. The low-frequency sounds emitted by the newly introduced boat traffic associated with the port is thought to have served as an attractant for sharks (Hazin *et al.* 2013). In fact, the prevalence rate of unprovoked shark bites has been directly linked to the number of ships entering Suape Port (Hazin *et al.* 2008). Several measures have been put in place in Recife to mitigate the risk of shark attack, and there has been (on average) only one attack per year in Brazil over the last 10 years, including just two fatalities.

Although not such a sudden change, an increase in shark bite incidence off the French territory of Réunion Island in the Indian Ocean, east of Madagascar, has also sparked interest and concern. While incidence rates are relatively low, they are slowly increasing and a large percentage of bites (66.6% in the last 10 years) are fatal (McPhee 2014). The small island is a popular tourist destination that promotes its pristine beaches, marine recreational activities (especially surfing) and wealth of marine life. Thus, the growing concern and publicity surrounding shark attack in this location has the capacity to threaten the island's visitation rates and, consequently, its economy. Interestingly, certain efforts related to protecting and preserving the beaches and marine life surrounding the island, and the related rise in population

I was heavily involved in investigating the possible causes of shark attack incidents and developing viable and ecologically balanced mitigating measures in Recife (Brazil) for many years. I was also the primary scientific representative to the Committee for the Monitoring of Shark Attack Incidents (CEMIT) from its inception.

There is still a lot of fear among beachgoers in Recife now. However, over the past decades there has been a clear increase in awareness and understanding that the sharks are there and that it's us invading their habitat. Therefore, it is us who should be careful when entering the sea. I believe that, gradually, people have blamed the sharks less and less for the incidents.

Moving forward, I feel that the most important direction for shark attack research is to develop ecologically balanced mitigation alternatives that are able to reduce shark attack incidents without harming the sharks. We also need to use the windows of opportunity these incidents create to educate people on how important sharks are for the balance of marine ecosystems and, therefore, how important it is for us to protect them. I feel the best approach to do this is through outreach and education. People need to understand that when they enter the sea, they assume the risk of being bitten by a shark, the same way they assume the risk of being bitten by a snake when they enter the forest. They also need to understand the precautions they should observe to avoid such incidents. Sharks are in their natural habitat in the sea, in their home. It's us invading their space, not the other way around. Sharks are magnificent creatures crucial to the balance of marine ecosystem, yet many species are threatened and facing the danger of extinction because of human action. It is our duty and obligation to protect them.

This field has always been very challenging, but mainly because of the massive coverage it receives from the media. And whatever attracts media attention also attracts the attention of crazy and insane people. I have been attacked, slandered, defamed and threatened by some of these people representing both extremes. Some were angry because we were 'harming' the sharks, while others were angry because we were not harming the sharks, we were 'freeing' them and, therefore, harming human beings. I used to be consulted quite intensively after an attack to identify the species responsible or to provide details of the event. This was until a couple of years ago, when I decided to abandon entirely this line of research and stopped giving interviews on the subject. Media accounts in Brazil are heavily misrepresentative. The media have very little interest in conveying scientifically accurate information. They are much more interested in generating sensationalised, bloody headlines to sell more newspapers. This is their business, and good news is non-newsworthy.

Professor Fabio Hazin
Universidade Federal Rural de Pernambuco

and tourism based on the marketing of these environments, have been proposed as contributing to the rise in shark bite incidence. In an effort to maintain the marine population status quo, legislation was enacted to protect the local reefs in the form of large protected marine parks. Further legislation was passed to limit the catch and removal of sharks from the area; however, that was enacted to stem the potential for the spread of ciguatera. The breadth and participation rates in water and underwater sports around Réunion Island have also flourished, meaning that humans are in the water and sharing the sharks' environment more than before. Furthermore, Réunion Island is heavily reliant on agricultural exports, but runoff and wastewater from the industry are poorly contained (Marie and Rallu 2012). As a result, increased amounts of pesticides and nutrients are released into surrounding waterways, which can affect fish health and the trophic structure and function of the ecosystem.

Potentially dangerous species

More often than not, if you hear or read about a major shark attack, the news is accompanied by stock footage of a white shark leaping out of the water, biting at a cage or swimming directly at the camera, jaws agape. Occasionally, the footage is of a grey nurse shark (*Carcharias taurus*) instead, whose long, pointy, continuously bared teeth portray a perfect image of a ferocious, man-eating beast (even though this is characteristically not the case). White sharks, along with bull and tiger sharks, do account for the majority of identified species in attacks, and 56% of recent attacks on humans have been attributed to these three species collectively (McPhee 2014). White, tiger and bull sharks were equally responsible for all but three of the fatal attacks between 1982 and 2011 (the remaining three were attributed to oceanic whitetip sharks, *Carcharhinus longimanus*). Not only do the sheer size and anatomy of these species make them a formidable threat to any large animal in their vicinity, but their ecology amplifies their potential threat to humans. These species are cosmopolitan in distribution, and their widely global, temperate, subtropical and tropical ranges tend to overlap with many popular human tourism and recreational locations. Although

Bull sharks are one of the species most implicated in bites at Réunion Island, and have been the target of selective culls in recent years. Photo by Tyrone Canning.

white, tiger and bull sharks are heavily implicated in attacks on humans, practically any large shark of around 2 m or more realistically presents a potential threat to humans (although not even this is a defining factor for potential negative interaction). The cookiecutter shark (*Isistius brasiliensis*), as an interesting case in point, is a slow-moving, reclusive, bioluminescent dwarf shark that reaches a maximum of 56 cm (Castro 2011b). Yet there is one confirmed and documented case of a cookiecutter shark bite on a human (ISAF 2017), and others have been speculated. In addition to the three most notorious species, and the surprising case of the cookiecutter shark, 30 other shark species have also been implicated in bites on humans (ISAF 2017), equating to just under 7% of all shark species. However, many of these species have been implicated in only one or a few bites in recorded history.

The prevalence of species implication in shark attack data, however, must be taken in context. The data would be surely skewed towards the larger, more iconic species, which are more identifiable. In reality, it is likely that Carcharhinid (or requiem) sharks are responsible for far more attacks than those currently documented, and certain species in this family would represent some of the major assailants. However, for

The iconic white shark (*Carcharodon carcharias*) is one of the most notorious shark species on the planet, and is one of the top three species implicated in bites on humans. Photo by Denice Askebrink.

several reasons, these sharks are rarely identified. To contribute to shark attack statistics, the animal must be confidently identified by the person bitten, a witness, through bite assessment or the recovery and analysis of remnant tooth fragments. First, it cannot be expected that most people would know of, or have a taxonomic interest in, the potentially dozens of species that may be present in an area. Second, most attacks occur over a course of mere seconds. Thus, unless there are unique identifying characteristics of an attacking species (e.g. the unmistakable cephalophoil, or 'hammer', of a hammerhead shark, which is still not definitive of a single species), it is unlikely that a victim would have a chance to adequately assess (or even see) the attacking animal. Third, many closely related species are very similar in appearance. This especially holds true for Carcharhinid sharks. One highly relevant example is in Florida, where a large proportion of shark attacks occur. Most attacks there are attributed to blacktip, spinner (*Carcharhinus brevipinna*) and blacknose (*Carcharhinus acronotus*) sharks (ISAF 2017); however, these species are almost inseparable in appearance. Indeed, the taxonomical differences between these – and some other closely related Carcharhinid species – revolve around key identifying criteria, such as

Many Carcharhinid sharks, such as grey reef sharks (*Carcharhinus amblyrhynchos*, left) are very similar in appearance (and conform to the typical 'shark' image). As such, they can be extremely difficult to differentiate. Tiger sharks (*Galeocerdo cuvier*, right), however, are more easily distinguishable due to their highly characteristic squared-off snout and striped marking pattern. Photos by Tyrone Canning (left) and Denice Askebrink (right).

the relative location of fins compared to other anatomical features, minor variations in body girth and the smoothness of an individual edge of a particular tooth (although this applies only to animals over a certain size) (Castro 2011a). Clearly, these criteria would be impossible to distinguish unless the animals were immobilised and lined up next to each other, and an ample amount of time was available for assessment. Further compounding the issue of identification is that body colouration within these species can vary by region.

Potentially dangerous activities

There is no standard condition for when sharks bite, and incidents occur over a range of aquatic environments, times of day and victim activities. Shark attacks have been recorded on humans who have been doing just about everything imaginable in the water. The list of activities includes crossing a river; cleaning/repairing/maintaining ships; bathing; falling off boats, jetties and piers; falling from planes, hot air balloons, etc.; performing a rescue of another human, shark or other animal; defecating in the water; playing with dogs in the water; fly-fishing; performing lifesaving drills; horseback riding; sleeping on a boat (which a shark reportedly jumped into, resulting in a fatal injury); and washing clothes, dogs, cooking utensils, horses, fish, etc. (Shark Research Institute 2016). However, while many of these activities are represented by one or only a few incidents, certain activities have a much higher incidence rate, notably recreational scuba diving, fishing/spearfishing, surfing (or similar activities), and swimming, wading or playing in the water. The increased incidence rate during these activities is not surprising, given that the rate of participation (especially in locations that reliably report and record shark attack incidence) would be much higher than some of the more obscure activities.

Analyses of shark attack statistics provide some unique historical surprises. Alcatraz, which is located in San Francisco Bay, operated as a US federal prison for 29 years (1934–1963). During its tenure, 36 men were involved in 14 separate escape attempts (two men participated in two attempts, each) (Federal Bureau of Prisons 2016). Three of the escape attempts are listed in the Global Shark Attack File for possible shark bites in 1962. Two cases remain unconfirmed, with the men more likely to have drowned than been fatally bitten by a shark; however, the third case is supposedly confirmed, with the man reported to have had two of his toes bitten off by a shark during his escape. The body of a man wearing an outfit akin to that of an Alcatraz prisoner was found washed up a short distance along the San Francisco coast several weeks after an escape, but the body was too badly deteriorated to be identified and the cause of death could not be concluded. Paradoxically, the resulting rate of one confirmed shark bite during Alcatraz escape attempts was equivalent to the one non-fatal shark bite on a surfer over the entire US coast during the same 29-year period, despite surfing being a far more popular activity.

Surfing tends to be the activity most associated with shark attack risk, and the nature of surfing has been hypothesised as exposing people to sharks in a greater capacity than most other water activities. This is attributed to the environmental and preferred conditions that are fundamental to the sport, including the surfers' duration of time in the water, their distance from shore, the often isolated activity locations and the 'provocative' arm and leg movements employed during surfing (Burgess *et al.* 2010). Thus, it could be inferred that beaches or locations that have the natural (or artificial) aesthetic, infrastructural and oceanographic characteristics that attract surfers would show increased prevalence in shark–human interaction and appear to be 'riskier' locales; this is indeed the case for certain Florida beaches.

Swimming and diving likewise present risk factors in relation to shark attack. Although quantifiable data on the subject are difficult to obtain, swimming and its inclusive activities of wading, playing and standing in the water, would likely include the largest percentage of recreational water users, due to being (generally) free of charge, easily accessible and requiring minimal equipment, training or resources. Thus, it is logical that an increase in the likelihood of interaction with sharks during this activity would occur, based purely on numbers. This assumption is supported by a study that surveyed Perth (Australia) beach usage and found that the most common activity of recreational beach users, over all beaches surveyed, was swimming; snorkelling, surfing and fishing were also listed as prominent activities on certain beaches (Eliot *et al.* 2005). Consequently, it is not surprising that swimming comprises a high percentage of shark attack incidents. It is also not surprising that only 14.7% of incidents during swimming and related activities proved fatal. Although only marginally so, this percentage is lower than for snorkelling, spearfishing or scuba diving (McPhee 2014), with the latter activities generally taking participants further off-shore and to more remote/isolated locations (lessening the crucial ability to obtain assistance rapidly). Although certainly not the rule, larger, potentially more dangerous sharks are often found further off-shore and thus away from swimmers. Although scuba diving would attract fewer participants than swimming, due to the associated cost, equipment requirements and qualifications needed (which correspond to lower participation and, consequently, incidence levels), the activity (similar to surfing) would place its participants further off-shore, in more remote areas, away from greater human presence and further from advanced medical assistance (corresponding to a higher percentage of fatalities).

Other aquatic activities also involve inherent risks. Sharks have developed an incredible ability to sense their environment and to locate food sources (as discussed previously). Thus, introducing an easily detectable, highly favourable (i.e. injured or debilitated) prey item into an environment that sharks frequent is naturally going to increase the likelihood of a close encounter, and this is precisely what many fishing activities do. Spearfishing, in particular, puts the participant at risk. A speared, bleeding fish will send out a suite of signals to predators:

chemosensory signals (blood and hormones), palpable mechanosensory, electrical and ampullary distress signals (increased, erratic movements) and potentially audible distress tones, which are quickly dispersed through the water. Responding predators may quickly become aggressive to win the meal, and the fishers are attached (via spear or line) to the shark's prey. Spearfishers are fully submerged in the water and have no easy retreat, so are particularly at risk. Other fishing modalities also place participants at risk, but as they often take place less completely or extremely in the realm of sharks, the level of associated risk is correspondingly decreased.

Conclusion

Sharks have always been a relevant concern to humans who have used open waterways, and accounts of shark encounters have been passed on through traditional forms of communication. However, modern communication abilities have revolutionised the way we perceive shark attacks. Currently, 3.2 billion people can connect to the internet (ITU 2015). This represents a nearly seven-fold increase in internet spread, from 6.5% of the global population in 2000 to 43% of the population in 2015. The percentage of households with internet access increased from 18% in 2005 to 46% in 2015. Thus, shark attack news is not only readily accessible and overwhelmingly reported, it is also instantaneously global. People have more ways than ever to perpetuate conversations about shark attacks, such as through social media and by commenting on electronic publications.

Other notable differences in shark attacks similarly revolve around improved technology. As mentioned, the increase in human water usage undoubtedly increases the likelihood and prevalence of shark attack, and it is largely through better technology (e.g. wetsuit design, more efficient diving equipment) that humans are able to spend more time in the water, over wider expanses and all seasons. On the other hand, advanced technology leading to improved medical equipment and quicker response times for primary, secondary and tertiary medical care have been major contributors to the significant reduction in shark attack fatalities over the last century. And of course, increased knowledge of how to mitigate the risk of shark attacks, and how to respond if the situation does occur, have been important in reducing prevalence and fatalities. This again comes back to increased communication.

While it must be acknowledged that shark attacks do occur, it must also be stressed that sharks very rarely bite humans and that it is even more rare for those bites to be serious, let alone catastrophic. Overall, the impact of sharks on humans – as a species – has been minimal (Caldicott *et al.* 2001). For most of us, the realism of shark attacks relates only to how good a job the special effects teams and videographers have done in the latest Hollywood blockbusters or in recent episodes of *Shark Week*. For most, the fear of sharks is an inconsequential, residual fear from a time when our ancestors were more regularly subject to natural predators. Although, for those who do spend a fair bit of time in, on or under the water in various regions of the world, the potential risk posed by sharks is one to be considered. This is not to say that the fear should be overplayed or exaggerated, only that the risk should be recognised so that appropriate precautions are taken through increased knowledge, experience and good judgment.

Over the last decade, there have been an average of 77 unprovoked shark attacks across the globe each year. Less than 12% of those were fatal, and this number is decreasing rapidly. While shark attacks have become global news, the countries that regularly record shark attack incidence on an annual basis (or even in most years) can be counted on two hands. More than 90% of global bites regularly occur in only 10 countries/territories and, in some cases, incidents within those places are isolated to, or heavily focused in, very limited stretches of coastline (e.g. Volusia County and Recife).

It is undeniable that shark attack incidence across the globe is increasing, as is our perception of the frequency of the events. However, we are also starting to gain a better grasp of why this is happening. Increased shark bite incidence is predominantly attributed to greater human beach usage, but several other natural and anthropogenic factors are likely to be involved. Encouragingly, certain high-risk areas, such as Recife, have been successful in decreasing shark bite prevalence over recent years, due to effective mitigation measures and greater public education (Hazin and Afonso 2014).

4

The role of the media in shark attacks: the good, the bad and the ugly

I hate horror films, and simply do not understand the appeal of purposely putting yourself in a situation where you will be stressed, uncomfortable and terrified. Watching the evening news gives me enough things to worry about and be terrified of, so why would I pay to be subjected to an extended cut of horror? I remember going to see *Jurassic Park* in the cinema when I was a teenager, which is rated PG-13 for intense science fiction terror. Let's be honest, *Jurassic Park* is not a real 'horror' movie. But I sat through most of it with my eyes covered. That being said, I have watched *Jaws* (also rated PG-13 for intense sequences of disturbing violence, among other things) many times, and I have suffered through all the sequels to the original masterpiece. I can't remember how old I was when I first saw *Jaws*, but I did like it. Funnily enough, I was not overly scared by it. While I didn't have any lasting fright reactions to *Jaws*, my logic for not wishing to subject myself to terror and horror in films or on television is justified. In a study designed to assess the fright effects of scary movies, 96.4% of people reported being affected by some movie, at some stage. A staggering 26.1% of people reported that negative effects lasted at least one year and were still ongoing, and another 9.4% reported a duration of one year or more although the effects eventually terminated (Harrison and Cantor 1999). We're not talking about trifling effects, either; examples included an inability to sleep through the night for months after watching a movie, and an unwavering avoidance of the situations portrayed (even to unrealistic extents). Most (81.9%) participants reported feeling at least one symptom of anxiety, such as crying, screaming, trembling, shaking, nausea, stomach pain, increased heart rate, a feeling of paralysis, fear of losing control, sweating, chills, fear of dying, shortness of breath, dizziness, faintness or numbness. I think I'll continue to cover my eyes.

Jaws was an instant box-office hit upon its release in 1975, and any search will show that it is still considered one of the greatest films ever made. It was well received, and the breadth of the audience it attracted was extraordinary and far-reaching. Its effect has been devastating, however, in terms of shark conservation, the human fear of sharks and a rational understanding of the animals. Suddenly, people who had never given a prior thought to sharks, let alone had any sort of previous interaction with these animals, now had a visually imprinted, conscious reason to fear and hate them. In the eyes of the public, sharks became rogue man-eaters, with a vendetta against humans. Almost overnight, white sharks (*Carcharodon carcharias*) went from 'being considered – at most – an obscure ocean dweller that few had ever heard of to a man-eating monster with a lust for wanton killing, and a creature that was best eradicated from our planet forever' (Peschak 2006). Even now, more than 40 years after the release of the movie, parallels and references to it are still widely used in relation to shark attack events and, remarkably, even in policy-making. Shark horror movies did not stop there, either (why would they, when *Jaws* was so successful?). Indeed, it seems no other predator has been so repeatedly and consistently demonised by cinema.

A growing public interest in sharks and increased awareness of them has led to a proliferation of media interest, from Hollywood cinema and television documentaries to print and social media. Sharks are big business and no other animal, on land or in the water, generates the entertainment income that sharks do (Neff 2015). Sharks are scary, and the threat of shark attack alone 'can make even the most ridiculous movies, like *Sharknado*, a hit' (CNN 2013). We are drawn to sharks, whether as fictitious creatures (in films) or in documentaries, because there is so much that we don't know about them. We find it intriguing that despite all our technological advances and our mastery of so many aspects of our environment, there are still mysteries like this in our natural world – particularly ones that can be so dangerous to humans. In a sense, sharks are the last truly wild animal. And in this way, we are mesmerised by them.

The media is a huge industry, and its effects are far-reaching. As such, it acts as a steward for how we perceive issues in our society and environment and can be a great educational resource. We learn from the media and, consciously or not, what we see and hear through the media has great influence on our views and opinions, and has long-term impacts on public perception, awareness and common stigmas surrounding sharks. However, the media is often at fault for elevating underqualified people to 'expert' status, for presenting opinion as fact, and for perpetuating myths. It also often uses terminology, metaphors, catchphrases, depictions and visual images to influence our feelings on a topic or skew reality.

Media coverage of shark attacks

Shark attacks are major news events, and they are generally highly publicised. When scanning the results of a Google search on the most recent shark attack fatality in Australia, I stopped counting after the 100th story that focused on the one event. The articles appeared in at least four languages and were from national, regional and online publications stemming from four continents. Governmental shark attack mitigation measures and policies also receive a vast amount of attention. Over a five-month period between December 2013 and May 2014, following the announcement and implementation of drumline deployment in Western Australia, more than 1100 radio news bulletins, 850 talkback comments, 290 television news pieces and 765 newspaper articles referring to the program were produced in Western Australia alone (DPC 2014). There was a further onslaught of attention on social media sites, such as Facebook and Twitter.

McCagh *et al.* (2015) analysed the reporting of shark attacks by a single print and online publication, the *West Australian*, which reaches approximately one million readers each week. It was found that in the two-month period following each of the seven fatal shark attacks in Western Australia between August 2010 and April 2014, an average of 45 articles (and up to 73 articles) were reported in that one paper. A news editor told me that shark attacks are so commonly reported because they are such a popular topic (along with whales and other marine animals) and attract high public interest. But coverage of sharks is highly skewed towards attacks, while other topics are not nearly as highly publicised. Between 2000 and 2010, there were well over 5000 scientific journal articles published on sharks, whereas there were significantly fewer (647) shark attacks across the globe. Yet an analysis of 300 shark-related media articles over the same period covering 20 major Australian and US newspapers found that shark attacks were the primary topic of over half the articles (52%). Scientific research relating to shark biology or ecology was the focus of just 7% (Muter *et al.* 2013). Despite Australia and the US being two of the most proactive and influential countries in terms of global shark conservation, this topic was only emphasised in 11% of articles. To receptive audiences, this bias in media coverage would reinforce the idea that, although we continue to learn about sharks and many species of sharks are facing severe conservational issues, the most prevalent matter is the risk they pose to humans.

Terminology

The terminology around shark–human interaction is something I've thought a lot about. I've considered it heavily in regards to the words I've used in this book. Research tells me that I should not be perpetuating the sensationalism of the matter through the use of 'shark attack', but should stick to more accurate terms like 'shark bite'. But ultimately, I purposefully chose to use 'shark attack' in many cases throughout the text. This was a conscious decision, not taken lightly. First of all, I want to connect with a broad audience, and I believe that using the term 'shark attack' best positions me to do this. But I also personally feel that in some contexts, the term 'shark attack' is appropriate. There is a lot of repetition in the English language, and most words have many synonyms. While this is not necessarily the case with 'encounter', 'bite', 'attack' and 'mauling', for example, I do believe that there is some overlap. And while I have not been in the position myself, after listening to people's accounts of their interactions with sharks, there is no doubt that in some of those cases I would be using the term 'attack' as well. Don't get me wrong, most bites are just that – bites – but I don't think we can say that attacks don't occur. The delineation between terms can be complex, though. We commonly say snake bite, spider bite and mosquito bite, yet the number of fatalities that result from those bites are greater than those from sharks. Thus, the terminology can't be based solely on outcome. Furthermore, the image of a tapeworm 'attack' is laughable. But, according to common terminology, people are *attacked* by bears, crocodiles and sharks. So, is the determining factor the size of the opponent? Or the size of the teeth? Or perhaps it mirrors intent, with 'attack' suggesting predation and 'bite' implying defence. This doesn't hold true for parasites, though. We simply don't understand animal motivation enough to be able to use this as a defining factor, and certainly not with sharks.

Ironically, the concepts of 'man-eater' sharks, 'rogue' sharks and 'shark attacks' all originated from scientific studies. The classification of white sharks by Carl Linnaeus in 1758 suggested a member of this species was likely to be responsible for swallowing Jonah, an account that was highly publicised in 1679 (as referenced in Neff and Hueter 2013), and thus attributed a man-eating motivation to the species from first identification. The initial terminology for shark bites (including fatal bites) in Australian government reports was shark 'accidents' (Neff 2012). However, as reported incidence increased and international correspondence on the subject accumulated, the terminology began to take on a more motivated tone, with words like 'attack' and 'man killers' used in scientific reports by Sydney surgeon Sir Victor Coppleson (1933). Coppleson (1950, 1959) was also responsible for terming sharks as 'rogue', as he argued that the only sharks that bit humans were those that had developed a taste for human flesh (which remains scientifically unevidenced).

Subsequently, the terminology surrounding shark–human interaction became more and more sensationalised. A high degree of intent, motivation and criminalisation began to be attributed to shark bite incidents (and even to normal shark behaviour, in general). Verbs such as 'patrolling' or 'cruising', with malicious connotations that imply threat, were chosen over innocent but scientifically accurate ones such as 'swimming'. And, although there were no behavioural or circumstantial changes, instead of 'biting', sharks began 'molesting' and 'assaulting' people. While shark attacks would, without a doubt, be 'horrifying' in some cases, occasionally people really are 'mauled', and in very rare events sharks literally are 'man-eaters', horrific maulings by man-eaters represent only the smallest proportion of the already rare event of shark–human interaction. And it is yet to be proven that 'rogue' sharks exist anywhere except in horror films. Another debatable aspect of terminology stems from the concept of boundaries. What constitutes the human realm, and that of sharks? This should be obvious – we are terrestrial animals and they are aquatic animals. However, Peace (2015) suggests that there is a liminal zone of the coastline that supports both species. As such, sharks are often stated to be 'violating', 'encroaching on', 'intruding into', 'stalking', 'lurking', 'prowling', 'loitering' and 'marauding' in these vicinities; again ascribing negative connotations and criminalised, intent-laden terms to a natural and rarely negative shark behaviour – swimming in their natural environment.

A search of media articles found that language used in relation to fatal shark–human interaction was skewed towards emotive language, as opposed to prescriptive language (McCagh *et al.* 2015). Emotional terms (with the frequency used) included man-eater (11), rogue (13), shark attack (353), *Jaws* (eight), killer (53), monster (10) and horror (four), whereas the descriptive terms sighting (141), bite (67) and encounter (51) were used far less. The use of descriptive language over emotive language can quickly change the tone of a media report and can influence the audience's opinion of, or reaction to, a topic. While greater context would be necessary to draw more objective conclusions, in fairness, I don't think it is unreasonable to consider fatal shark bites to be 'attacks' or 'horrific'. These terms, however, would not be as justified for many non-fatal bites, yet they are often still used in this context. The use of emotive language for effect is also often exploited by politicians when promoting new mitigation policies. They generally refer to sharks as 'rogue' or 'man-eaters', giving the animals the punishable human villain-like characteristics of intent and motivation (Neff and Hueter 2013).

A search of Associated Press articles in Florida newspapers with the keyword 'shark attack' highlighted 48 articles between 8 July and 25 August 2001, the period corresponding to the infamous 'Summer of the Shark' (Neff and Hueter 2013). Although no fatalities resulted from those incidents, the word 'attack' was used 201 times, or once every 159 words, demonstrating a real discord between actual events and media reporting. A recommendation was made for the alternate use of four more realistic and descriptive outcome-based labels: shark sighting, shark encounter (physical contact between sharks and humans or their equipment, with no resulting injuries), shark bite (resulting in minor to major injuries) and fatal shark bite. These terms remove implied intent, represent the diversity of the topic and provide a more objective depiction of the relationship between sharks and humans. The authors suggested that 'shark attack' should only be used where intent can be clearly established, after assessment by qualified experts, to say that the interaction was a direct result of predation or defence. However, the authors argued that our current lack of knowledge of the physical, chemical and biological triggers that lead sharks to bite humans realistically precludes such assessment.

Ultimately, I feel that the wide range of circumstances surrounding shark–human encounters deserves a similar diversity of terminology. I would argue that shark attacks are rare, even in the grand scheme of negative shark–human interaction, but do occur. Really, in terms of terminology, I think that considered judgment and a focus on providing the most accurate assessment of the situation is what is necessary. There are over 170 000 words in the English language, let's start using more of them to provide realistic accounts of these situations.

The bad and the ugly

The biggest issues relating to media representation of shark bites are the disproportionate overabundance and sensationalism often afforded to these topics. Humans overweigh the probability of rare events when they are easily related to and recalled. Therefore, these attributes of the

It's a very hard gig, journalism, and I think we've probably got to cut journalists some slack. Some writing sounds gratuitous, but if you think you hear clichés, that's largely because journalists need to be writing several articles a day now in most cases. You can't be reinventing the wheel every time, so sometimes you reach for a well-worn phrase. With my recent piece on sharks and shark attacks, 'High Hazard', I was given the scope and the time to work on it properly and to gather the necessary knowledge, which was great. But most journalists aren't routinely given that kind of time, even if they do desire to write responsibly and well, which most do.

In my experience, most journalists don't think about the nuance of terminology much, if at all, when writing a story; they're not really taught to do so. It's sad, because language is so incredibly powerful. But you have to consider that there are all sorts of nuances in terminology for all sorts of issues. There's nuance in disability terminology, there's nuance in race terminology, there's nuance in scientific articles. And generally, the only people who appreciate the nuance of terms at all are the people who are deeply involved in those kinds of subjects. A disability advocate would know what terminology is preferable to people with disabilities, but a journalist wouldn't. A journalist relies on being told and being trained, and there are inadequate resources to train a journalist in everything they might conceivably need to know. People are quick to whinge about whatever the media's done now, but there's not a lot of reaching out to journalists to inform them about what would be better to say. So, in most cases, journalists simply don't realise that they might not be using best-practice terminology. There's an assumption that journalists are malicious when they may be just unaware. When I spoke to a Fisheries representative while researching shark attacks, she told me that Fisheries have select terms that they use: 'encounter', 'buzzing', 'serious shark bite', or 'fatal shark bite'. They only use these terms, and 'attack' is never used. The word 'attack' is just a quick go-to term the media use. It wouldn't even occur to most of us to consider using other words. Not everyone has the knowledge about what constitutes an attack versus an encounter, or that the two should even be differentiated. Not to mention that 'Serious Shark Bite' is just not as compelling a headline as 'Shark Attack', which is clear, concise and sounds dramatic. Of course we write headlines for maximum attention grab. But I do believe that nuance is important, and now that I understand the situation a bit more, and especially now that I have actually been informed about the different terms, I'm certainly going to try to increase my use of nuanced terminology. But I also think that it would be optimistic in the extreme to assume that I will influence anyone else by doing so. It will be more a personal principle. I'm sure my headlines will continue to say 'attack', though, because that's what works. And I must say, even though the media can't know about every nuance that an interest group might push in order to be more precise, eventually we do learn what is considered appropriate, and what is not. But we won't ever learn if scientists and government officials and interest groups don't push to have distinctions recognised. If there is a particular reason to say 'serious shark bite' rather than 'shark attack', it is really the responsibility of these people to make their case heard. People need to be a bit more forthright in saying, 'You can rephrase this, because I know it's a bit boring and it doesn't matter if you paraphrase, but don't paraphrase this because it shows an important distinction.' If they don't, journalists will of course continue to say 'shark attack, shark attack, shark attack'. Just as with any group of people, the media industry contains good people and bad people. And just as with any industry, most journalists care about their work and want to do it well. Trust me, they're not there for big money, adoring public, cushy workload or job security!

With a major news event, especially, you can't afford to waste any time. And every shark attack is a major news event in Australian media. Whether it's a bite, fatal, or even just an encounter, it's big. So, as a journalist, you can't afford to waste even minutes, because if your competition gets that news onto social media before you do, then you're going to lose vital traffic, and traffic means advertising, which means jobs. You have to verify the basic information as quickly as you can and publish immediately. Often, you first hear about an attack through social media from people who are at the scene. You'd generally then get that initial information verified by the Department of Fisheries. And then it's a matter of just getting that basic information up, even if it's just a few lines early on. Everything else is then refined and expanded on as events unfold, so you'd continue adding to that core story for several hours, days, or possibly even weeks, or eventually add new stories around that initial news story for as long as readership numbers indicate that people are interested.

Choosing who to speak to and quote in a story comes very much down to individual journalists and who their contacts are. Since I've now written a few stories on sharks, I have some people on file that I can call. Generally, a journalist would call the media liaison for the Department of Fisheries and maybe the Minister for Environment's office, as well. But otherwise, it all depends on how good your contacts are. If you don't have any contacts who may be able to comment from an individual viewpoint, you'd consider what organisations or interest groups there are that are potentially affected by these things. So, for shark attacks, you'd potentially be contacting organisations like Surf Lifesaving Western Australia and asking them for a comment, or St John Ambulance, if they were involved. And you'd start with people who have commented before, or who you think might comment again. And again, in those early stages, you're just searching for people who can provide context around what you need. In terms of initial immediate coverage of a shark attack, you're not really thinking about framing a debate, so who you contact for a comment, based on their expected viewpoint, would not be a consideration. It's different for follow-ups to the initial story (those that attempt to address the 'how and why', as well as the 'who, what, where') or long, research-based stories, where you'd be a lot more conscious about attempting to balance a variety of viewpoints. But you don't really have that luxury if you're just reporting on a shark attack, you just need whoever you can get on the phone quickly. And later, if people contact you and say, 'I have something to add', then you'd consider taking the story further depending on whether that expert can add new context or understanding for readers. Bias will always be a problem. Publications can have agendas, and it is clear that not everybody cares purely about the science. But bias is clear to an educated reader. We must hope readers consume a variety of news sources, but the increasing homogenisation of news producers and companies does not bode well in terms of the public being able to access a range of viewpoints, especially with social media algorithms feeding them news they already agree with and lessening the opportunity for people to have their pre-existing beliefs challenged.

Sharks are definitely a topic of interest. A lot of the stuff that I was reading in the research, and that I agree with, revolves around that ancient primal narrative of man's mastery of the unknown. I think that that's also why shark hunting and shark fishing are so prevalent; they are a way to prove superiority and assert control over something uncontrollable – the ocean and what's in it. But when there's been an attack, you sense a desire for revenge, not always from the victims or their families, but from the public at large; the desire to regain a sense of control and mastery. And then of course there is just that fascination from things like *Jaws* and the movie culture. I think that runs really deep in people. But I agree that the media does fuel this fire among the public with repeated stories on shark attacks. Absolutely. And it doesn't surprise me that there are 40-something articles per shark attack, but you have to think about it as the chicken and the egg kind of thing. People don't like clickbait, but then they click on it. People don't like articles that 'beat up' shark attacks, but then they click on all the articles. Media outlets respond to the traffic that they get, and if we see that the shark attack story is rating strongly, we're going to keep writing about it in response to the demand. Media corporations cannot afford to ignore traffic. If people were really against these kind of stories, then they would stop reading them. And then we would stop writing them.

Emma Young
Fairfax Media

media inappropriately reinforce the occurrence and hazard of human–shark interaction.

Misrepresentation of true animal behaviour in movies such as *Jaws* portray sharks as 'rational enemies', resolved on attacking swimmers (Papson 1992; Neff 2012). False links between animals' actions and their intended purposes ('teleological fallacies') appear in films, television and nature documentaries alike, although to varying degrees (Stone 1989). Indeed, nature documentaries have been described as 'less a window through which we watch nature, than a reflection of value-laden culturally defined perceptions' (Papson 1992). While sensationalism may be useful for gaining traction and interest, it can also perpetuate misunderstanding in audiences who believe they are being shown unstaged footage of natural behaviour. For example, to incite enough activity to show shark feeding behaviour, documentary producers may unnaturally draw sharks into an area through heavy baiting or chumming of the water, leading to a feeding frenzy. Thus, although the resulting footage would show feeding behaviour, the setting, scale and timeframe of the situation may be highly uncharacteristic of the animals filmed. The need to lure sharks to a particular location for filming presents an additional concern in terms of shark attack. These activities have the potential to increase shark presence, activity and feeding response in the vicinity. This could present a heightened risk for other water users in the surrounding area who may not be aware of the increased shark presence and activity. Although only anecdotally reported, isolated accounts of this have been recounted to me by several people in relation to various locations.

While there is much potential for good in media reporting on sharks, it unfortunately tends to be outweighed by the negative. The Discovery Channel's *Shark Week* is a major annual media event and has the potential to inform and educate a large, captivated and wide-reaching audience. Yet the conservation and research messages of these shows are often overpowered by the presentation of the animals' potential violence. This trend is replicated throughout various media sources. In their analysis of shark-related newspapers articles, Muter *et al.* (2013) found that more than half (59%) emphasised negative effects such as human bites, deaths and closed beaches, whereas only 19% focused on the positive effects (e.g. shark-inspired human medical breakthroughs). The remaining 22% focused on positive or negative effects on sharks (e.g. increasing or decreasing shark population trends).

Film and television

Despite the identified purpose of *Shark Week* being to dispel myths and misconceptions about sharks, lessen the fear of sharks and enhance the overall knowledge of sharks, the shows simultaneously promote fear through the use of vivid imagery and description. There is a disproportionate reliance on the threat of shark attack to draw in and engage viewers. Shark attacks are consistently referred to; shark attack victims are interviewed, providing graphic descriptions (and often photographs) of their interactions and wounds; and viewers are regularly reminded of the risk of negative interaction with sharks through the narrative. Television viewers import attitudes on risk of violence from what they watch (Gerbner 1969; Morgan and Shanahan 2010) and even a single exposure to shark–human violence could elicit feelings of fear, because this concept is suddenly accessible in a person's mind (Bushman 1998; Myrick and Evans 2014). People are prone to negative attitude changes (e.g. increased dislike or avoidance of something) following media representation of related frightful situations (Harrison and Cantor 1999).

Three types of fear-arousing stimuli are commonly used in film and television media to elicit fright reactions in audiences: portrayal of danger or injuries, injury-related bodily mutations, and the dangers and fears experienced by others, all of which are common in media representations of shark attacks (Myrick and Evans 2014). The shot used most frequently during the entire 10-documentary series of *Shark Week 90* showed a white shark breaking the water's surface, rolling its eyes back into its head and ripping a piece of bait apart. This single shot, undeniably aimed at fuelling the human fear of being attacked by a shark, appeared 45 times throughout the series (Papson 1992). The in-programming promos for the series included statements such as 'Cross the fin line of terror all weekend long. Waters off the Australian coast seem peaceful, but just beneath the surface there lurks a menace', 'The smell of blood in the water, a magnetic attraction to these man-eaters', 'When they're hungry, and that's all the time, no creature is safe. Their senses are keen, their jaws deadly' and 'This tornado of terror indicates one thing: a feeding frenzy. Behold the shark's insatiable appetite ... But first travel Australia's treacherous seas searching for the man-eating monster, the great white'. These promos included fear-invoking background music and heavy intonation by the narrator. Yet, during the series, viewers were told that they should not let irrational fears of sharks dominate their thoughts.

While imagery is a main driver in fear-eliciting response, background music is also influential. After viewing a 60-second video clip of sharks swimming, with ominous background music, participants who had watched the otherwise uneventful video rated sharks more negatively than those who watched the same video set to uplifting background music, or in silence (Nosal *et al.* 2016). The notoriously ominous music in *Jaws* was identified as a particularly frightening part of the movie, and to some it was even more frightening than the presence of the shark, or blood (Harrison and Cantor 1999).

Print and electronic media

Despite its high potential, print and electronic media regularly fails to meet the educational and informational needs of readers, instead focusing on accounts of negative interactions. A survey of Western Australian residents showed that 56% of metropolitan and 72% of regional residents wanted to know what beaches were closed as a result of shark sightings or activity. More than half of the people surveyed also wanted to know how to personally reduce the likelihood of encountering a shark, information about the behaviour of sharks in their region, and the facts and figures of shark sightings and movement. Fewer people were interested in what the government was doing to reduce the risk of negative shark–human interaction, or the facts and figures about the true incidence rate of those events. These results are in direct contrast to many media stories, which most often focus on the number of attacks in a given timeframe or governmental action, while research on shark behaviour and movement often fails to attract significant media interest.

A really ugly side to shark attack media has recently emerged; one that can severely compromise the mental health of people who have been affected by negative shark encounters. Cyberbullying/harassment and shark attacks seem like completely distant and unrelated topics, so I was surprised to learn of the extent of the link between them. I was first alerted to the fact that many online publications have a comments section where people can discuss the posted stories when a colleague suggested that I check what had been written in response to an article that discussed his recent research findings. One stand-out comment ranted that the researcher had been paid off by sharks and was taking kickbacks from them. This type of comment is laughable, not really fitting the definition of bullying, and is of little real concern (except about the mental stability of the person making the comment). However, more serious threats were also directed at the researcher. He didn't feel that any of them warranted sufficient concern, but they were threats nonetheless. Unfortunately, this was not an isolated incident, and many other accounts are far more disturbing.

Sharks and shark attack mitigation are such controversial and highly mediated topics that bite victims and their families and friends may quickly be caught up in, and have their behaviour analysed in relation to, tangential issues raised by the media following attacks. This could relate to shark culls, shark status or new mitigation measures. A lot of public anger can come out in relation to these issues, and the association between the victim and the anger can be really confronting for people who are already dealing with the mental and physical burdens of trauma. Many media pieces focus on the survival techniques of the person bitten, which is great for public awareness and knowledge, but it can concurrently de-emphasise the horror that survivors would have gone through. Various interest groups may also take the opportunity to capitalise on the newsworthiness of shark attack events to push their own agenda, disregarding the trauma of the experience and feelings of the people involved.

It turns out, amazingly, that many shark attack victims are openly abused through online forums for their audacity to have been bitten by a shark. While most media outputs are extremely respectful to the individuals involved, and to their family, friends and others affected, the readers who comment on related media articles and through social media are occasionally not. People post hateful, abusive comments about the person bitten (despite not knowing them) and criticism of their behaviour (e.g. swimming, surfing or diving) because those activities exposed them to sharks. Not only are these comments socially inappropriate, but they can contribute to overwhelming feelings of isolation and dejection in shark attack victims.

Shark attacks – the media and tourism

Thanks to increased communication abilities and modern media, shark attacks are no longer localised occurrences. Regardless of their origin, individual shark attack incidents are often reported by media in many different countries. Thus, it is not illogical to think that this sort of information would be taken into consideration by prospective tourists to a region, as well as by local residents. The effects of shark attacks on tourism are regularly cited as a reason for increased regional shark attack mitigation measures. However, the correlation between the perceived risk of shark attack and tourism statistics is inconclusive.

A website designed to support the hire and rental of various products, Hiresquare.com.au, conducted a survey of people who had previously shown an interest in coming to Australia for a holiday from the five most important countries for Australian tourism (China, the UK, the US, New Zealand and Japan) (Hiresquare 2015). It concluded that, following extensive media coverage of Australian shark attacks, 6.1% of respondents no longer wanted to come to Australia due to the risk of shark attack and a further 7.8% said shark attacks significantly impacted their decision. The potential 14% loss on tourism equated to an impact of AU$2.9 billion in revenue. Interestingly, data from this study showed that of the 42 respondents interested in visiting the Australian Capital Territory, 14% stated that they would not go anymore, 33% stated significant or little doubt, and 52% stated that shark attack had no impact on their decision. Similarly, of the 40 respondents interested in visiting the Northern Territory, 13% stated that they would not go anymore, 36% stated significant or little doubt, and 53% stated that shark attack had no impact on their decision. Ironically, the percentage of people who stated that shark attack had no impact on their decision was the *lowest* for these two places over all other states and territories in Australia. Yet, the Australian Capital Territory is well and truly landlocked, has never had a single recorded shark attack and is not even acknowledged in the Australian Shark Attack File. Although it has a significant stretch of coastline, the Northern Territory has only one recorded unprovoked shark attack since 1990 (in 2005, when a spearfisher received minor cuts) and just

12 (confirmed and questionable) cases in recorded history (Shark Research Institute 2016). These places represent the lowest incidence states/territories in Australia by far. These results poignantly demonstrate the amazing discord between reality and public perception.

The correlation of shark attack statistics and beach usage, however, fails to confirm a link between the two (Banks 2015). Numerous examples can be found showing that regional beach attendance (as recorded by the Australian Lifeguard Service, Surf Life Saving New South Wales and New South Wales Tourism) increased despite shark attacks in the area, or fell regardless of a lack of attacks. Communication with the Research Office at Tourism Western Australia stated that there was no current evidence, anecdotal or otherwise, that had considered the impacts of shark hazards and mitigation on tourism. However, it was noted that the natural beauty, coastlines and beaches of Western Australia were the top motivators used in the tourism industry, which could imply that potential and past travellers remain positive about Western Australia's beaches, despite potential shark risk.

The good

Despite all the negatives, the media can be a powerful and positive resource in relation to sharks. It has been identified that people want to be better informed on sharks and shark attacks, and the media is a preferred and easily accessible source of information. Being aware of any potentially hazardous situation is always beneficial, and knowledge is power. While it is true that an overrepresentation of shark attack stories can lead to false perception and increased fear among the public, a realistic acknowledgment of the hazard is a good thing; in this regard, the media is an informant's strongest ally.

Sharks characteristically draw interest. Thus, well-researched, objectively written media pieces on shark attacks can be highly beneficial in educating broad audiences. The most beneficial stories alert people to the risk while also drawing attention to the rarity of attacks, presenting preventative measures to reduce human vulnerability to attacks, promoting responsive action and addressing relevant conservation and research issues. Repeated proactive communication about ways to reduce the risk of encountering or being bitten by a shark could help to reduce incidence rates and negative perceptions of sharks.

Media resources are a valuable communication tool for both immediate and long-term hazard information. If a potentially dangerous shark has been spotted, or increased shark activity is suspected for a given area, the use of various media modalities to rapidly disseminate that information is highly beneficial to public safety. While signage and lifesavers can notify people in a limited vicinity, many beaches are unpatrolled and unsigned. A broader, more widespread means of information dissemination has the potential to quickly reach a much larger and more remotely populated audience. The Western Australian Government developed an integrated shark notification and response system in collaboration with the Department of Fisheries, Western Australia Water Police and Surf Life Saving Western Australia that provides real-time information on shark sightings and detection to response agencies and the public via Twitter and the Sharksmart website. When an acoustically tagged shark swims within 500 m of a receiver buoy (or when a shark sighting is reported to the authorities), the information (including species, location and date/time) is automatically posted. This system allows people to make informed decisions about their water activities.

In stark contrast to the significant damage that social media can cause those affected by shark attacks, when used proactively, it can be equally beneficial to victims' mental health. As stated repeatedly, shark bites are incredibly rare and their incidence is spread across the globe. The effects of the events on those involved can be highly traumatic, life-changing and incredibly isolating. However, the advent of social media has allowed a few highly motivated and highly perceptive people to begin to turn this around. The Bite Club Facebook page was started in 2011 with the purpose of forming a support network for those directly affected by shark attack. The core aim, according to information on the group's page, is 'to assist those affected by shark attacks in returning to a "new normal" life, and to support and assist them throughout the journey'. While initially designed to support a local group of survivors in New South Wales (Australia), the group now comprises more than 250 people

Shark attacks affect more than just the person bitten. Every guy that was in the water with me that afternoon was also a victim of my shark attack. They all played a role, and it had a mental impact on all of them. And even good mates of mine who usually surf with us, but who weren't with us that afternoon, didn't surf for up to six months afterwards, they were so shocked and freaked out about what had happened. And then you have your family, the paramedics, who I met sometime afterwards, and the helicopter pilot and the crew who had attended to me. They would have also been affected by it. I think I put the number of people who were directly related to my shark attack at 20. Thinking through that really surprised me, and this is now one of the things that we're looking at in Bite Club, and something that we're trying to bring up with the government, to create some sort of action plan that can be used if somebody is attacked.

There are a range of different opinions on sharks among the members of Bite Club, and we encourage these differing views in order to achieve a realistic balance within the group. My opinion is that any shark attack survivor, regardless of their views, will always be welcome in this group, and in my house, because they've all been through the trauma that brought us together. So as far as the views that survivors have go, I can speak with anyone that I've met through the shark attack world, and I can get where each of them are coming from. Some of them want to kill every shark, some want to go into shark conservation. The very reactions to how people think about sharks afterwards are as diverse as our backgrounds were to start off. So I insist that personal views are not an issue among the group; as a group, we stand for one thing, and that is to look after the humans that have been affected by shark attack. Another guy said it beautifully once, there are literally thousands of groups that look after sharks, but there is only one that looks after the people who have been attacked by sharks, and that is us. There are other shark attack survivor groups, but from what I have seen, they don't have the same priorities.

One of the reasons we started the Bite Club group in 2011 is because people on social media can be real, for want of a better term, bastards. What happened to me, which literally blew my mind, was there I was, laying in hospital hours after my attack, and I started reading some of the newspaper stories on the computer about the attack. For some reason, my attack made the newspapers all over Australia. I hadn't realised previously that there were comments sections at the end of these articles, but I happened to see them this time, and I started reading them. I was shocked to find that the comments were mostly directed at me. There was all of this anger towards me, because I had been attacked by a shark. The comments were basically saying, 'I guess he now wants to go out and kill all of the sharks because he's been attacked', and, 'What does this idiot think he's doing swimming in the dark'. But just to clarify, and to provide a bit of background on the circumstances, it definitely wasn't dark at the time of my attack.

We live in a somewhat remote location, and there aren't a huge number of ambulances here. And incidentally, there had also been a car accident the same afternoon, and all of the paramedics had been called out to it. So it was more than an hour after my attack that paramedics turned up. The paramedics then worked on me for about an hour and a half on the beach to try to stabilise my heart. So the photos that were taken of me while I was being loaded into the ambulance – the ones that appeared in these newspaper articles – were taken after dark, and that's where a lot of the comments about what I was doing surfing after dark came from. But in reality, these photos had been taken nearly three hours after the attack, when I was actually transported from the beach.

I was so surprised by what I was reading in the comments that I started answering them then and there, from my hospital bed. One of my first comments was, 'I take full responsibility for using the ocean and what happens in the ocean'. But the abusive and negative comments just kept coming. I went on for four hours, on the newspaper sites, trying to defend the fact that not only did I accept the fact of being attacked by a shark in the ocean, but also that I did not to one degree want revenge on the shark that had attacked me. But not one person acknowledged these things, because they already had their minds made up. I became so infuriated that the last comment that I posted on one of these news sites ended up being, 'You have no idea what my life is about and what I am about, so go and f*** yourself'. And that was about how I finished it. When I looked into some of the people posting later on, I found that they were nearly all related to a single conservation group.

Once I started looking a bit further, I found that it was regularly the same sorts of people, making the same sorts of comments towards anyone who was attacked, and these types of comments just kept coming up on social media. So I continued to fight back for the people who had been attacked, and I used to make comments like, 'If you were at the beach when this person was attacked, would you run around saying "It's your own fault, it's your own fault" or would you lend a hand to help this other human being?' And also, 'If it were one of your loved ones, would you be saying the same things that you're saying now?' And these comments usually defused a lot of the smart comments after that. I've found that if you can make people put themselves in the situation, then you can usually get people to think about how they would react if it were someone they knew on the beach bleeding to death and needing help. But these comments on social media and in the press still appear after every attack, and it has really blown me away. The conservation groups are ripping into shark attack survivors, just for having the audacity to have gotten attacked in the first place. And this has often been the topic that has come up first up in conversation when I meet with other shark attack victims for the first time. People say to me, 'I can't understand where the anger towards me came from. I was a good bloke when I went into the water, and when I got out, I'm no longer that same person; I'm instead the enemy of everyone'. And that's how you feel; a persecuted minority is one way to put it. There becomes a lot of anger towards you after your shark attack and the way you get treated. And to make it worse, you're mentally compromised from the attack itself, you feel alone, you can't get answers from the Department of Primary Industries, you can't get answers from scientific organisations, you can't get answers from doctors sometimes, and everyone wants to hate you, except for your own family, but they're going through the same thing. So this is something that we're trying to turn around and fix in Australia.

The hardest part of my attack was the isolation. There was no one who came and sat with me in hospital and said, 'Hey, Dave, this is what you can expect'. And nobody from mental health came and saw me, and said, 'You've just been through a major traumatic event', and talked to me about it. Because of nearly dying on the beach (I lost 40% of my blood as a result of the attack), and then obviously not knowing what the future was going to bring, I was isolated. Friends came to say 'Hi' and ask how I was doing, and wish me the best, which was awesome in the hospital, but then I left hospital and went home. And both my daughter and partner, who were living with me, had to go back to work, because I wasn't going to be able to work for quite a while and we still had bills to pay. So I was discharged from hospital, less than a week after my shark attack, with a whole pile of pain medication and basically told 'Off you go, sort yourself out'. I got regular visits from the community nurse to change bandages and dressings and such, but it was the time alone, when I closed my eyes, that I really struggled. I would have nightmares of recollections of what had happened that afternoon. I would wake up screaming, and that was something that I had never, ever experienced before in my life.

I reached out to many people after my shark attack to see if there was anyone who could answer some of the questions that I had, and basically just hit walls everywhere. So I started finding other shark attack victims in my area, and started reaching out to them. But in my search for answers, what I actually found was not answers, as such, but instead I found a whole heap of people who were feeling the same way I was, who had nobody to relate to, nobody to answer their questions, and nobody who was even remotely interested in trying to answer their questions. I started meeting these people, and as soon as we sat down and started discussing some of the things that I had been through, they opened up about going through the exact same things. So when the next shark attack happened, which wasn't long after mine, I got in touch with the hospital, and said, 'Hey, I'll just leave my contact details, have this guy ring me, and I can help him with what he might be going through based on what I went through'. And that's how this started. It's not only helped other people by just having that initial chat, but it also reconfirmed that what I had been through was a normal reaction, and we were able to work on it together. So from that one little search of mine, we've ended up with over 300 people who now regularly communicate with each other from all over the world. Over the last 12 months, I've probably talked to six new shark attack survivors. And over the last five years, I've probably spoken to over 100 shark attack survivors. I think it's important to try to help these people to deal with what's ahead of them, and I encourage others to do the same. There are several people that I've met that have been really supportive of others, especially ones who are now missing limbs and such. It's something that I'd never planned on, but it's become really important in my life.

Although I did struggle mentally at first, I got it down fairly quickly after talking to some of the others. There was a girl who was attacked the same week as me, and I met her the day I got out of hospital. I would ring her quite often to discuss some of the things that we were both going through. Her being quite young, the same age as my daughter, it was the worst thing that had ever happened to her. For me, being much older and carrying around a lot more baggage that took precedence in the personal trauma department, the injuries I received and the trauma of the shark attack itself, probably didn't make the top five worst things in my life. I figured that I could just look back on the other times that I had had to deal with mental trauma, and how I got through those times, and then just apply the same thinking to my shark attack. So I thought originally that I would have had a really fast bounce-back from the shark attack, but I didn't. It was just so different. There were a few things in particular that I was experiencing, which obviously others facing debilitating traumas would have also had to deal with, but it was mostly just the fact that I had no one to relate to over my shark attack.

There have been several people in our group who have talked about contemplating suicide. They just haven't been able to cope at all at some specific time. Luckily, we haven't lost anyone to suicide, but unfortunately we've had quite a few who have come close to it. Because like I've said, it's sometimes very hard for someone to find the right people to relate to. And if you can't find anyone to relate to, then the isolation of being on your own and thinking that no matter what you do, you can't find anyone to talk to, or to make this better for you, is enough to make people think about just ending everything. And luckily, a few of those people, at least, have made great turn-arounds. So we started up another Facebook page a while ago called Beyond the Bite, and basically it's the most effective of all our groups. Bite Club includes shark attack survivors, rescuers and their families, who also often have an interest in what's happening and want to ask questions and things. But Beyond the Bite is about moving forward, and it's for asking the questions that you're not game enough to ask anybody else. Like 'Hey, guys, I'm having a really bad day here, I am not handling things very well' and that sort of stuff. There are physical injury questions, because we all have various stages of disablement and there are things that some of us can't do anymore, especially for those who have lost a limb. This is not only frustrating, but also depressing for some people. So we've also taken that group to the next level, and we're now a registered charity. We're aiming to secure funds for mental health treatment for those within our group who really need it, and to raise funds for people to just get back to a new normal life. And by this, we refer to a reality that will never be the same as what it was, but it can be a new normal for them. So if we have somebody who needs a new leg, then we can help them to get that leg. It's now all about not only supporting each other with barbecues and someone to talk to and a big hug, but also supporting people financially to start their new lives, so that they can get themselves back on track to where they need to be. Shark attacks can be very expensive, as well as physically and mentally debilitating.

One of the saddest things I've seen from shark attack survivors, and for me, having been a surfer for 40-odd years, is you see people give up the ocean. There are plenty of surfing advertisements that say only a surfer knows the feeling and the love for the ocean, and I can relate to that. The passion is there forever, and to have to give that away, through mainly fear, would be very devastating for somebody's life and mental health. I still surf as much as I always have, but there is one saying that I have used several times, 'I miss the blissful ignorance that I had before my shark attack'. We all knew there were sharks out there, we would see them from time to time, but it was good to be ignorant of what would happen to you if you were to be attacked. I miss those days. Since my shark attack, what I have noticed, and I guess what I was starting to notice before the attack as well, was the increasing number of sharks that we have in the area. Now I have a thirst for knowledge to find out what the hell is going on. I had six close encounters in the year after my attack. And I don't know if I had some sort of smell on me that attracted them, or if I had developed an extra sense that I could see them more, but one of these encounters nearly ended my surfing. I got out straightaway, threw my board on the beach and I said, 'That's it, I'm not surfing anymore', and I proceeded to sit down on the beach and cry. I was that sad that it was over. It was like giving up part of my life. It was a very emotional moment. But luckily, my partner and my son were both there, and they came and sat down with me, and we got through that together. But to have that feeling, and to feel like I had no option but to give it all up, was quite disturbing to me. I'm now back in the ocean full-time, like I had been previously, but I am very acutely aware of what is happening in the environment. If the water is a bit smelly, or looks a bit fishy, or if I notice birds diving, or anything that indicates that there may be sharks around, then I just can't surf.

While I'm glad to be back in the water now, I am still faced with the dilemma that whenever I've had issues in my life previously, I'd go surfing. But now, the place I used to go to sort out life's problems is actually the source of one of those problems. And what had previously been my place of solace is now sometimes my place of fear. All these years later, it is still a struggle, and it's still emotionally difficult. I've met guys recently that were attacked by a shark 40-odd years ago, but they can still describe the attack as if it had happened yesterday, down to basically every detail. And pretty much every shark attack survivor that I have spoken to can do that, because it becomes so ingrained in your brain – the whole event and how terrifying it was, that 40 years later, people can still speak of the event as if it were yesterday. I actually wrote mine down as a story so that I wouldn't forget anything, but I've never bothered to read it, because I've never actually forgotten anything.

We are looking into all types of shark deterrents – Sharkbanz, Shark Shields. But personally, I'm just being more observant. I take the responsibility of surfing in the ocean personally, just like you would when you chose to swim in a pool. I know the risks of going into the ocean, and even more so now. I take my own safety into importance. I'm not fully convinced of the effectiveness of any of the current personal mitigation measures. And unfortunately, at least half of the attacks that I've spoken to people about have involved the shark coming in at high speed. Mine hit me, and flipped not only me but the shark actually flipped itself out of the water as well, like you see them do on TV. And mine was a bull shark, not a white shark. They can apparently be quite aggressive when they decide there is something in the distance that they want to have a crack at, and they just make up their mind and come in. In those cases, I don't believe that anything would deter them. Fortunately, in my case, my head just happened to hit the shark on the nose when it hit me, and that's probably one of the best things you can do to a shark, is hurt it on the nose. It still dragged me under the water with it, and we had a bit of a cuddle, then parted ways. I was able to get back to the surface and onto my board, and somehow I got through it. But the shark continued to swim around underneath me for quite a while. I guess it was just my way of dealing with it, but I got through that part of the situation by thinking that the shark was most likely just swimming around, wondering what the hell had just happened to it, while I was sitting there assessing the damage, wondering what the hell had just happened to me. I'm not sure that I would have been able to handle sitting there if I was thinking 'Hang on, it's going to come back and have another go at me'. And luckily, I was soon able to focus a bit better and started concentrating on getting out of the ocean. I yelled out to my mates to get out of the water because the shark was hanging around, but two of them made the decision to stay and make sure that I was all right first. I had thought that I was on my own, though, and tried to paddle back to the beach, but just ended up getting cleaned up by the waves. After nearly drowning, I came back to the surface. Luckily my mates were there, and they helped me to get back onto my board again and get out of the ocean, which apparently took quite a while according to the people on the beach.

I'd like to stress that despite some pretty tough times, particularly right after my attack, things have not all been doom and gloom. My life is awesome. The way I see it, the day of my attack was the worst day for me, and also the best day of my life. I nearly died, but I didn't. And I choose to look at it from the perspective that I didn't die. So that's something really positive that I got from my shark attack; I've got a whole new appreciation of life now. I'm the most positive person in any room, at any given time. Shark attacks can have that effect on people. I have a great life now, probably better than what I had previously, and I now don't let drama get in the way of anything. I no longer accept all of the drama and whinging, and if people whinge to me now, I just say, 'Come on, I know guys who have to go to the toilet, and they don't have any hands anymore. You think you've got problems?'

Dave Pearson

Bite Club Facebook page (open only to those affected by shark attack, who need support and who wish to proactively help others to cope with what has happened):
https://www.facebook.com/groups/501367279913000/
Contact: biteclubaustralia@gmail.com

around the world and has been highly beneficial to those struggling to come to terms with their experiences. In particular, the increased connectivity has reduced the severe feelings of isolation. As isolation is commonly linked with post-traumatic stress disorder, the camaraderie provided by the group may lessen the degree of that burden within the shark attack community. Members with different points of view on the topic are welcomed, in order to encourage a successful balance.

Social media allows people who have experienced similar situations to more easily contact and reach out to new shark bite victims to offer support. As shark attacks are such singular experiences, which humans are not psychologically prepared for, this sort of support and understanding in the early days following an attack has been identified as integral to survivors' recovery, helping them to realise that they will be all right, that there are options available to them and that they will still be able to live their lives and accomplish their goals.

People who have not been in the same situation simply don't understand, because they haven't lived through what you have lived through. And this is a point where we could quite interestingly contrast being a bushfire survivor in a small town to being a shark attack survivor. One is shared, whereas the other is very singular experience. However, the literature suggests that a sense of lacking social support following trauma is a predictor of who goes on to have more difficulty in terms of recovery. So that's a really important point when thinking about how to look after yourself. Others may not understand, may react by glossing over, or may continue asking about the event, all of which could lead to distressing memories. So, it is really understandable why people shy away from company, but it's still really important for recovery to stay connected and engaged with people, even if they may not be able to give you the specific support that you need. One of the symptoms of post-traumatic stress disorder (PTSD) is avoidance, and part of that is social avoidance. People pulling away from relationships because of anxiety and fear is common in PTSD, and that's one of the reasons why people might have difficulty recovering, because they will avoid social interaction and social support.

Dr Richard Cash
Senior Clinical Specialist, Centre for Posttraumatic Mental Health, Department of Psychiatry, The University of Melbourne

Conclusion

The value of sharks in the entertainment industry is undeniable, and there are both pros and cons to the intensive media representation that sharks draw. Ultimately, what is needed is balance. Ideally, sensationalised films that tap into the deep evolutionary fears of humans should be counteracted with realistic representations of sharks' natural behaviours and accounts of non-negative shark–human interaction in documentaries and through real-life experiences (e.g. visiting public aquariums). Interestingly, while *Jaws* was a groundbreaking media event, none of the (numerous) subsequent films that portrayed shark attacks have had anywhere near the success of *Jaws*. Unless, that is, you consider *Finding Nemo* or *Shark Tale* to be 'shark attack' movies. Perhaps this is because shark attacks are no longer novel or unheard of; they have become common in living rooms around the world. News coverage of shark attacks should certainly continue; however, these pieces should be balanced in terms of providing the details necessary to inform the public, while also including factual information on the rarity of such events and proactive insight into how to mitigate risks. The sheer quantity of stories on shark attacks should be decreased, while accurate stories on shark research and shark attack mitigation should be increased. The use of non-sensationalised terminology, appropriate to the situation, should be used.

Promisingly, changes in media do appear to be occurring. Analysis of print media has suggested that articles focusing on the negative effects of sharks have decreased slightly, but significantly, over time, and those discussing the positive effects of sharks have increased (Muter *et al.* 2013). Advances in social media have allowed for more proactive and wide-reaching shark conservation campaigns that help to raise public awareness on sharks in general. And finally, despite the large degree of sensationalism and many inaccuracies in film, television and print media, the media undeniably feeds the public's fascination with sharks. And this is a good thing! The sociobiologist E.O. Wilson once said, 'We're not just afraid of predators, we're transfixed by them, prone to weave stories and fables and chatter endlessly about them, because fascination creates preparedness, and preparedness, survival. In a deeply tribal sense, we love our monsters.' We protect, support and look after the things that fascinate us, we make efforts to learn more about them and we seek more experiences with them. So, while the information and narrative from current media sources may be false or exaggerated, anything that inspires people to learn more about sharks is a bonus. And this holds especially true in today's society, where accurate, factual information can easily be acquired if we make the effort to look for it. Finally, greater familiarity with sharks may limit panic situations.

5

The fear of the improbable: human psychology and shark attack

Martin, it's all psychological. You yell 'Barracuda!', everybody says, 'Huh? What?' You yell 'Shark!', we've got a panic on our hands on the Fourth of July.

Mayor Vaughn, *Jaws*

How better to start a chapter on the human fear of sharks than with a very appropriate, and surprisingly perceptive, quote from the movie *Jaws*. But when the statistical prevalence of shark attack is so low, exactly why are humans so scared of sharks, and why do so many of us have such a consciously unjustified fear of being attacked by a shark?

One of the first things that we must understand is that this seemingly irrational fear of shark attacks is not unique. According to some accounts, the prevalence of human fear in general is thought to be increasing. The German sociologist Ulrich Beck coined the term 'risk society' in 1986 to describe the heightened level of risk that he predicts society is headed towards. He suggested that people are on the verge of being frightened unlike ever before due to the global shift of human life and the environment we occupy (Beck 1992). However, the fear of sharks is nothing new.

While I have become slightly afraid of sharks in certain circumstances (a nasty side effect of thinking about shark attacks on a daily basis), I do not have a strong or restrictive fear of these animals. Generally, when asked if I am afraid of sharks, I respond with the clichéd but terminologically accurate response that I have a 'healthy respect' for these animals. Despite its commonality, I like this phrase and now, analysing it a bit further with my mind on psychology and medicine, I fully appreciate the word 'healthy'. It adequately serves the purpose of, by definition, asserting that I do not have a clinical level of fear or anxiety of sharks. But a 'healthy' respect also means that I wish to protect myself and keep myself safe. I think both of these things are important. And a 'respect' for the animals, instead of fear, suggests understanding rather than irrationality. I know very well what these animals are capable of, but I also understand that their actions are purely biologically and physiologically driven. Of course that understanding won't protect me if I ever find myself in a negative encounter with a shark, but I can at least recognise that there is no malice or negative motivation in its actions. Sharks do not terrorise humans with an intent to instil fear; they simply do what they need to do to survive. And I absolutely respect these animals for what they are, and for what they have achieved.

However, I acknowledge that many people do not share my feelings towards sharks, nor my healthy respect rather than fear. Many people, even those who have never seen a shark and have no intention of entering the water, are afraid of sharks. I grew up oblivious to sharks. We would go to the beach for a week every summer, and I would play in the ocean any chance I got. The thought of a shark never entered my mind. In fact, I am not sure that I would have even known what a shark was at that stage. I can't say for certain if this represents a significant change in the media or if I just wasn't watching the right channels, but I don't remember even once hearing about a shark attack in the news during my childhood. My parents certainly never seemed concerned about sharks or prevented me from going into the water. In fact, my first memory of sharks was watching an episode of *Shark Week* that informed me that sharks were simply misunderstood, and were, in fact, vulnerable. The supporting footage showed a white shark (*Carcharodon carcharias*) rolling its eye back within the socket, seconds before attempting to making contact with a seal. In this case, as portrayed by the show, the seal, with its protective and destructive claws, was the villain – the viewer was encouraged to cheer on the shark. Many people will have had a different introduction to sharks. But my experience does help to explain my mindset compared to that of others, at least to an extent.

Fear and anxiety

Fear is a fundamental and complex emotion that is most appropriately explained with an understanding that it is inherently defiant of logic and rational thought. Both the origins and the maintenance of fears have proven to be more complex than was assumed by early models. More contemporary research has begun to account for the complexities through the consideration of individual vulnerability, invulnerability and context. And this helps us to understand how discrete events shape human fear, how anxiety disorders develop, and the most effective treatments for these conditions (Mineka and Zinbarg 2006). While fear appears to be illogical and defy conscious common sense, its importance in human evolution cannot be understated and it is considered to be highly relevant in the preservation of mammalian life over the last 8 million years.

Only a few studies have tried to analyse the human fear of sharks, specifically, and the results failed to produce any sort of characteristic data or trends. Consequently, there are no published scientific reports on the human fear of sharks. However, animal phobias are widely studied and generally elicit a conserved response signature. Instead of sharks, snakes (and, to a lesser degree, spiders) are the prototypical models used to study the psychological fear of potentially dangerous animals. Based on these models, it appears that one of the more explanatory theories of why some humans develop a seemingly illogical fear of sharks is because of a concept known as preparedness. This concept forms the basis of the evolutionary theory of fear development.

The evolutionary theory is based on the foundation that we are more prone to fear objects and circumstances that would have threatened the survival of our early ancestors. When considering human history, modern humans in modern conditions (i.e. the last few centuries) have been in existence for just a quarter of a hundredth of a per cent of human existence. That's 0.0025%. Thus, it is argued that the human brain was shaped by a world completely foreign to what we know today; risks were not the same, and survival was tested by a completely different set of factors. To survive in an environment fraught with predators, the development and implementation of effective predatory defence systems was necessary and this vital need is theorised to be the origin of mammalian fear (Arrindell *et al.* 1991; Öhman and Mineka 2001). These ancestral fears are generally biological in nature and reflect natural objects relevant to the survival of the species, hence the elevated role of animals (predators) in human fear. Similarly derived fears include heights or unfamiliar places, open or confined spaces, situations where normal sensory capabilities are compromised (e.g. dark places), and bodily deformity or disability. While the theory does not suggest that humans should fear sharks specifically, the particular identification of predators is not considered essential, nor even suggested. The sheer multitude of potential predators that would have been relevant to our human ancestors has resulted in learned, generalised adaptations to eluding these threats (Öhman *et al.* 1985). The most forward-thinking of our ancestors also learned to develop fears based on the experiences and perceptions of others (Mineka 1992; Olsson and Phelps 2007).

Throughout evolution, danger has had the potential to strike fast and without notice. Thus, an essential characteristic of fear is that it provokes avoidance and escape without reliance on the physiologically timely process of consulting with the conscious brain (Epstein 1972; Öhman and Mineka 2001). As such, a critical component of the evolutionary theory is that it operates and is acted on automatically, with minimal conscious or cognitive thought (Öhman and Mineka 2001).

Although evolutionarily based, prepared fears should not be interpreted as inborn fears but as those that are comparatively easy to acquire or particularly difficult to treat (Mineka and Zinbarg 2006). To account for the modern environment and contemporary risks, it is important to note that this theory does not preclude the development of fears that early humans would not have faced (e.g. airplanes, cars, chemical warfare); however, it does suggest that we don't develop a fear of these things as easily and that these fears are more easily forgotten. This contributes to the explanation of why people may be more afraid of shark attacks (which they will likely never experience) than of car accidents, which a large percentage of us are susceptible to on a daily basis, and have often had personal experiences with.

The evolutionary theory of fear acquisition finds further support in the (conserved) anatomical machinery and physiological response elicited by fear. Indeed, the neural circuitry responsible for fear and fear learning is shared among the entire mammalian lineage (Davis 1992; Kapp *et al.* 1992; Fanselow 1994; Davis and Lee 1998). The subcortical location of this behaviour (around the amygdala) suggests an ancient evolutionary origin stemming from animals with brains far more primitive than those of modern mammals. The anatomical location of the fear-processing centre suggests an unconscious, autonomic response, and it is believed that fear behaviour is not under voluntary control. As a result, we can sense fear even without consciously recognising something scary. It has been stated that, while fear is a conscious experience, the processing of fear, and fear assessment, are as unconscious and as automatic as the functioning of our organs (Rosen and Schulkin 1998).

Another popular theory of fear acquisition is the learning theory. This theory helps to answer why people are afraid of sharks and fear the possibility of shark attack when they have had no prior exposure to, or negative interaction with, such things. The learning theory incorporates the concept of conditioning, which revolves around the pairing of a neutral stimulus (e.g. shark) with a fear or pain-producing state (e.g. trauma, distress). This association turns something previously insignificant into something to be feared (Rachman 1977). The degree of fear a person feels to a stimulus can be related to the number of associations between the pain/fear experience and the stimulus, and the intensity of the adverse reaction during the association with the stimulus. Negative reactions to a stimulus do not need to be personally, or even realistically, experienced. Conditioned fears may be transferred from person to person, such as from parent to child, or even through the re-enactment or portrayal of someone else's fear response in a movie or the news (Cook and Mineka 1990; Mineka and Zinbarg 2006).

Yet, despite all the prepared opportunities to fear sharks, not everyone does. As fear is learned, a large part of our capacity to resist certain fears stems from our childhood or early experiences. Children raised by non-phobic or non-anxious parents (in relation to specific stimuli) show a reduction in the amount of fear or anxiety towards those stimuli, even if they personally have a negative experience with those things. To this extent, repeated observations of non-fearful behaviour to a particular stimulus can effectively immunise a child towards later fear acquisition (Lubow 1998; Mineka and Zinbarg 2006). Furthermore, research has found that children raised with more control over their environment are less frightened by, and are more able to manage, new frightening situations in general (Chorpita and Barlow 1998; Mineka and Zinbarg 2006).

Ultimately, it appears that the human fear of sharks has multiple bases. An overcautious fear of predators appears to be deeply seated (both anatomically and evolutionarily) in the human brain. Thus, this statistically illogical fear may actually be inherently logical. Second, we live in a world of information, and we are constantly learning. We learn not only from our parents, teachers and peers, but also from the media, politicians and organisations or groups that we associate with. A single movie that portrays shark attacks, involving in-depth character development and the character's repeated enactment of fearful behaviour, has the potential to spark fear in those who have never before given a thought to sharks.

Risk perception

I love going to the beach, snorkelling and diving, but I am careful where I choose to do these activities. As an informed person, with a respectable knowledge on the topic, I know just how slim my chances of encountering a shark are (which I actually would welcome), let alone having a negative encounter (which I do not). Yet there are times that I simply cannot shake the feeling of fear, and quickly abandon those activities.

Most contemporary theories of cognitive psychology and neuroscience revolve around the belief that humans utilise a two-part complementary system to perceive risk. The first is the analytical system, which uses developed algorithms and mathematical rules, including probability laws, risk assessments and logic (Slovic *et al.* 2004). The second is the experiential system, which uses intuition and is influenced by images and associations experientially linked to emotion and feeling. Both systems have advantages, limitations and biases. The analytical system is slow, requires effort and relies on conscious thought. Conversely, the experiential system is fast, predominantly automatic and largely independent of and unresponsive to conscious thought. While our conscious mind will often tell us that our feelings of risky situations are irrational, the rapidity and the emotion elicited by the experiential system generally means that this system wins. Thus, risk perception is more often than not based on feelings, rather than on conscious or rational thought. The experiential system is the most natural way humans assess risk. It is often described as gut feeling, sense or instinct. It is thought to be the way that our evolutionary ancestors survived, and it is what we still use today when we're under pressure.

As with the preparedness theory of fear acquisition, the experiential system of risk assessment has evolutionary origins in the human brain. This is in contrast to analytical thought, which arose much later in human evolution, to contend with the increasing complexity of the world. In terms of risk, thanks to the experiential system, if a situation (real or perceived, personal or communicated, fact or fiction) can be easily recalled, then that risk is perceived as common and relevant. And since emotional, vivid or unusual memories are longer-lasting, they are often more easily recalled.

Our ability to rationally perceive the risk of shark attack is further limited by the oversimplification of numbers by the human brain. It has been suggested that innumeracy is the natural human condition, and it takes significant effort to become numerate (Dehaene 2008). Humans easily recognise big and small numbers, but when numbers become really big or really small, then we lose the ability to assign meaning to them. We are less sensitive to proportionate changes in big numbers; for example, the difference between 0 and 1 is highly appreciable, whereas the difference between 800 and 900 is less so. This human inability is exploited by lotteries around the world, which rely on the fact that we simply cannot comprehend exactly how small a chance we have of winning. It appears that we compensate for this inability to truly comprehend really large or really small numbers by shifting our focus from probability (likely or unlikely) to possibility (yes or no) (Loewenstein *et al.* 2001). We also cope by turning numbers into accounts or stories. Risky situations are no longer statistical but become black and white – we are safe or unsafe – and the shades of grey in between are lost. The high improbability of winning the lottery is replaced with a simple yes, it's possible. And, although even more improbable, yes, it's possible that we may have a negative interaction with a shark if we cross into an environment that they inhabit. And while it is very unlikely that a person could recount the exact number of shark attacks that occurred in a year, many people would be able to recall hearing or reading about an attack. The numbers are irrelevant – what is important to the human brain is that these events do occur. We know this because we can recall hearing about such an event happening.

The lack of probability perception is further exacerbated by emotion, and this has been termed 'probability blindness'. If taking a chance on something could result in strong emotion (positive or negative), then the degree of attractiveness or unattractiveness of that thing is relatively insensitive to change in probability (Rottenstreich and Hsee 2001). Although shark attacks occur so rarely, they carry a high degree of negative emotion; therefore, the true probability of these events occurring is relatively meaningless. Instead we focus on the fact that they occur, and the extent of the negative consequences that could result.

While the discord between actual and perceived risk is inconsequential in some cases, in others, disproportionate risk (either over- or underestimated) can lead to significant consequences. If underestimated, failure to take ameliorative action could occur. However, if overestimated, a government may feel forced to respond to subjective public perception, for example, and unjustly apportion resources to less significant, less statistically worthy, concerns. In the case of shark attacks, insufficient response could result in apathy and a lack of preventative behaviours

It's hard to know where to start. I had known Doreen for a long time, and for that whole time she had been a diver. We'd had the conversation several times where she told me that she felt that if her time was up, then it was up, but I don't think that we'd ever had the conversation about sharks, specifically. She was very practical about the possibility that she could be badly injured, or that she could lose her life at any time doing any of the various things that she loved, but this never stopped her. Not that it was any less of a shock or tragedy when we lost her. And I can understand continuing to pursue activities that you love, despite a known risk. My husband and son are both avid motorbike riders, and there is a very real possibility that one of them may not come home one day, so we've had that type of conversation before as a family, as well. But still, there is no getting away from the fact that the loss has left a gaping hole in our small community.

As far as we understand, it was a lone shark that was in the area. And we just can't help but think what were the chances of her being in that exact spot? I mean, the odds were so low. Realistically, she'd have been more likely to have won the lottery that day. But one way or another, she's gone. And for all intents and purposes, it was because of an inanimate object. I know the shark was not actually inanimate, but it's no different to a motorbike. You can't blame the bike; there's no anger around that. There's just a gaping hole where that person was. And it was so sudden, and that was the main thing. If we had lost her to a lingering illness, I don't know if it would have been any easier, but it would have been different. But the fact that it was a shark doesn't make any difference to me.

I've always had to defend going into the ocean to my friends and family in the United Kingdom because they are terrified of the situation here. The news of major shark attacks in Australia makes it over to the UK. And there is always the sensationalism about sharks in Australia in the English news. If you go to the beach here in Western Australia in the summer, you see spotter planes, or notice that certain beaches are closed at times. I don't keep a close eye on the situation, but in my peripheral vision I have noticed that that sort of activity has been increasing. My own thoughts towards sharks have not changed, though. The ocean is their domain, we enter it at our own risk, and sometimes that means that we are no more than food, if that's how you want to look at it. I've got no anger towards the shark, or sharks in particular. There is no malice in their behaviour, it is just what they do. But I do think that there are ways that we could mitigate the risks against ourselves so that we can coexist more peacefully. They should be able to do what they need to do, while we are still able to enjoy the things we like to do recreationally.

I feel very strongly on the situation with sharks because there is another living creature here. And no one creature is better than another, and everyone should come out intact, not one or the other. This isn't my area of expertise at all, and not something I think about often, but if I've got any feelings towards risk reduction strategies, it would just be that everyone comes out unharmed.

There are a lot of common myths and misunderstanding around shark behaviour, and I think that's because there is not a lot known about it. But if there is emerging research, then we need to be getting that out there, so that people can be using that information. I have heard when there have been sharks in the area and when beaches have been closed, but to be honest I don't know how I knew those things. It must be through news bulletins on the radio or television. Whatever they are doing now, it's obviously working, because I am aware of it without actively seeking the information. But not everyone listens to the radio or the news, so it should also be all over Facebook and Twitter, and maybe it is. It just needs to be one short, sharp message about the current risk of shark attack that is broadcast as widely as possible to reach all audiences. And then it would just become common that you listen out for it, and routine. I think that people would take notice of that. Or at least 99% of people would. But you'll always get that 1% that doesn't, and there is nothing that you can do about that. The urban myth that you hear around the place is that there is a widespread view that overfishing is happening, and so that is bringing sharks closer to where recreational diving and other activities are happening. And I feel that that is something that really needs to be looked at.

I'm not a strong swimmer, so I probably only go to about neck height when swimming in the ocean, and I will probably still continue to do that. More so, though, I am a kayaker, and I will probably now think very carefully about kayaking out to sea. I might go along the coastline, but I wouldn't go out to sea. That's probably the only way that I have, or will, modify my behaviour. It's just what you're prepared to accept. I'll just choose to go to the river instead of the ocean, or just stay along the coast. I can accept modifying my kayaking route, whereas my husband and son can't accept giving up riding their motorbikes. You can't stop people from doing what they love. And really, you just wouldn't do anything if you were alerted to the risks of the situation all the time.

My family and my friends all feel very strongly about the drumlines, and all of that. We really don't feel that it's the right thing to do. It's the human supremacy thing that I hate. It's damaging what is really an innocent animal by labelling it as some sort of malicious evil; they're not, they're just animals doing what they are supposed to do. There must be other ways of detecting them and keeping us safe. I've got quite a few friends who are triathletes or competitive swimmers, and they train in the ocean. But now, even the ones who live up in the northern suburbs travel up to 80 km to train in the areas that have the shark exclusion nets. So there does seem to be some sort of recognition for that sort of technology, which allows them to train in a safe spot and not be distracted by the thought that they could be attacked at any minute. It seems to be a really effective measure. Kids can go in there and swim safely, competitive swimmers can go in and train, stand-up-paddleboarders go in there, it attracts all sorts of recreational water users. And I have found it really interesting that people who live on the coast already are choosing to travel so far to go down there. There's obviously some sort of behaviour modification happening. The response to the exclusion net was almost instantaneous when it was installed, so people were obviously holding out for something like that.

There is not a lot of conversation in public health care about post-traumatic stress disorder and preparing people to live with that or manage it. We've talked about it since Doreen's death, about adding something back into our curriculum for better preparing nurses and midwives. We talk about emotion and grief in terms of death, but we don't talk about it in terms of having to adapt to something lost, like where people survive a trauma, but still lose something very significant, like an arm or a leg. They would still experience a grief response, or a trauma response, and I don't know that we really do enough in preparing nurses to recognise or deal with that. We certainly prepare them well to deal with people who have lost somebody, full stop, but not when people survive an attack but their lives have been changed. I think that following this experience, we might make a point to add that back into our curriculum next year.

Friend of Doreen Collyer, who was fatally bitten by a shark on 5 June 2016, Perth, Western Australia

regarding further incidents, whereas an overreaction could lead to the unnecessary destruction of animals, avoidance of recreational water activities and unjustifiable resource allocation and policy to address the matter. There are documented examples of all of these responses.

Certain hazards display greater fluctuation and more inappropriate levels of public concern; these are termed 'panic'. In this context, panic is defined as a situation where rises in the level of public concern for a problem become either temporarily out of proportion with the problem's objective magnitude, or incongruent with past and future levels of concern (Loewenstein and Mather 1990). In panic situations, surges in concern do not match changes in the underlying problem and, during peaks, the level of concern is grossly out of line with normal levels. The proximate causes of panic are often particularly vivid events or an overwhelming amount of media attention. Inherently, panic events cannot be explained by objective changes in the underlying problem. Public panic may be met with overreactions by governments, an occurrence known as 'action bias' (Sunstein and Zeckhauser 2008). This concept is very relevant to shark attack, and large-scale panic situations have the potential to lead to poorly considered reactive measures, which can be detrimental if executed at high levels.

Conclusion

The root of the human fear of sharks is deep-seated in our subconscious brain. It can take as little as one negative experience with a shark, or even a media representation of someone else having a negative experience, for this fear to be realised. Thus, it is not surprising that the fear of sharks is so prevalent, even in locations where the risk of interacting with a shark is negligible. As a species that was once prey to a variety of animals, humans have evolved a variety of mechanisms that have saved lives, and, consequently, have been passed from generation to generation. These mechanisms mean that we are designed to sense potential danger rapidly and subconsciously. Despite the later development of the ability to consciously rationalise risk, this slow, evolutionarily modern system generally can't compete, and is often overruled by feelings. The downside in our inability to rationalise risk is that it can lead to unnecessary fear, public panic and misguided political action.

6

How to lessen the risk of shark attack: personal mitigation strategies

Aside from not entering the water, the best way to avoid a negative interaction with a shark is just to be smart about the situation and know the facts. Although shark bite incidence is already extremely low, there are several simple things that a person can do to further reduce their risk of having a negative encounter with a shark. For some, personal mitigation devices may be effective in reducing the fear and/or occurrence of a shark attack.

Know the facts and stay up to date

Education is probably one of the most effective shark attack mitigation methods. Taking the time to get to know sharks and form a greater understanding of these animals, as well as the facts around shark attacks, is probably the most important thing a person can do to mitigate the fear, likelihood and effects of shark attack. This sort of education should start as early as possible, but it is never too late to start learning. You don't need to learn to love sharks, but it is important to gain an understanding of their biology and behaviour. Taking the time to observe them carrying out their normal behaviours (most often simply swimming around) is a great first step and provides a refreshing dose of reality that contrasts with what we tend to see of sharks through movies and the media (where, more often than not, sharks are determinedly eating, attacking or stalking people). The easiest, safest, most controlled and most informative way to do this is to visit an aquarium. For those who want a bit more adventure with their education, there are a multitude of places where people can swim, snorkel or dive with a variety of species of sharks (from small benthic or reef sharks to large, potentially dangerous species) on their own or through commercial ventures. The more familiar a person becomes with something that could be considered fear-inducing (such as sharks), through any sort of positive or neutral experiences, the less they will perceive that object as a risk (Slovic *et al.* 1978). Familiarity also assists with post-encounter recovery; those who are more familiar with sharks are able to cope with encounters (positive, neutral or negative) far better than those who are not. This is most likely why many surfers, who often see or at least acknowledge the presence of sharks on a regular basis, are able to get back in the water soon after an attack, whereas family or friends who spend less time in the water may be unwilling or unable to do so. Regular exposure to sharks, through aquarium visits, personal education or encounters in the wild, can also help to strengthen a person's understanding of the actual threat that sharks pose.

Regardless of a person's background experience or knowledge of sharks, anyone who recreationally enters the water should take the time to investigate and have an understanding of the environment. This is no different from what should be done when going into any new environment (whether it be one with significant animal or environmental risks, strong cultural differences or an elevated crime rate). Taking this precaution before entering a natural waterway protects a person from far more than just sharks. For example, before swimming at the beach, it is always advisable to understand the water movements in the area. We seem to have 'domesticated' beaches for tourism and recreational purposes and appear to have forgotten, or ignore, the fact that they are not benign environments. They are wild, natural environments, they can be dynamic and unsafe, and they contain a suite of biological and physical features that have the potential to present significant risks.

A savvy, safe swimmer will look at the movement of the water, watch for dangerous currents and note benthic structures (e.g. sandbanks or drop-offs). These can all influence the dynamics of the environment, and the level of skill, strength and expertise a person needs to safely swim or take part in other water-based activities at a particular area. Similarly, it is always advisable to know the local rules and regulations that relate to beach and waterway usage. There may be bans or restrictions on the type of activities that can be undertaken in certain areas (e.g. swimming, surfing, motorised and non-motorised vessel usage). Also, it is wise to know the protocols for adhering to regional lifesaver policies, such as swimming only at monitored beaches, or only between the flags that indicate a supervised area. And of course, it is advisable to know how to contact lifeguards or other emergency responders in case something does go wrong. Most aquatic environments present some sort of biological risk, usually animal-related. This could be jellyfish, cone shells or sharks. Knowing the biological diversity of the area (with some understanding of relevant risk mitigation measures and the related first-aid if interaction does occur) can prepare a person for potential negative interaction and can also be of use to others. Hazards are often depicted on regional or beach-specific signage in high-use areas, and this information should be read and considered. However, such information may not always be in place, especially in more regional or isolated locations. Ultimately, it is up to the water user to know and understand the risks.

While we do not fully understand the movement patterns and motivations of sharks (and probably never will), continued research into this area will supplement our knowledge and provide a much better idea of when and why particular sharks might be in an area. For example, it has been found that the risk of shark attack on surfers in Mendocino (California) is 24 times lower in March than in October or November, and it is 1566 times safer to surf in the ocean between San Diego and Los Angeles (California) in March than it is in October or November in Mendocino (Ferretti *et al.* 2015). This sort of knowledge and information comes from research into the trends of shark behaviour and shark–human interaction. It provides the potential for highly relevant decreases in shark bite risk, well above those achieved through any other mitigation measure to date, including culling.

Therefore, keeping an ear and eye out for ongoing research on these topics is beneficial. Monitoring or early warning systems may also be highly useful. Both the Western Australia and New South Wales governments have websites and smartphone apps that alert people to the location of acoustically tagged sharks, and this technology will continue to grow and be refined. Accessing and utilising this information can help to inform decisions about when and where to enter the water. Other monitoring systems, like the Shark Spotters program in South Africa, also inform water users about the current risk of shark attack at a given beach.

Make wise choices

Although we cannot predict exactly when or where a shark attack will occur, we do know enough to know what sort of situations to avoid. Very generally, the best advice is to avoid areas where sharks are known to feed, conditions that would limit sensory capabilities and actions that may be representative of normal prey items. And, obviously, the use of basic common sense and intuition when in an environment with known predators is always beneficial.

According to the Australian Shark Attack File and the International Shark Attack File, the following measures should be employed to minimise potential negative interaction with a shark.

- Swim at patrolled beaches – the water is monitored and there are trained people there to assist if an interaction does occur.
- Do not conduct recreational water activities where potentially dangerous sharks are known to congregate.
- Always swim, dive or surf with other people (the presence of multiple people may deter a potential attack, and also a companion can assist you if a negative interaction does occur).
- Do not swim in dirty or turbid water where visibility would be reduced for both you and the shark.
- Avoid swimming at dusk, dawn or night (many sharks are more active during these times, and these times also afford low visibility).
- Avoid swimming far off-shore, near deep channels, between sandbars or along drop-offs to deeper water (sharks are more likely to inhabit deeper water, and being further off-shore places you further from assistance).

- Avoid entering the ocean near a river mouth, especially after a rainstorm (changes in salinity can affect shark behaviour and rain can wash potential food items into the sea, which might attract fish and sharks).
- Do not swim among schooling fish that have congregated in large numbers (which might attract feeding sharks), or where birds are diving (as this can indicate schools of baitfish).
- Do not swim near people fishing or spearfishing (these activities can attract sharks).
- Dolphins in the area do not indicate the absence of sharks (dolphins and sharks sometimes feed together and some sharks feed on dolphins).
- Kayakers should raft up together if a large shark is seen in the area (this creates the appearance of a larger object that a shark may not be interested in).
- Do not swim with pets and domestic animals (sharks can be attracted to the erratic splashing of non-aquatic animals in the water).
- Look carefully for the presence of sharks before jumping into the water from a boat or wharf.
- Be careful wading through shallow water as benthic sharks may be hiding there, and can accidentally be stepped on.
- Avoid wearing shiny jewellery in the water as it can reflect light, resembling the sheen of fish scales, and attract sharks.
- If a shark is sighted in the area, leave the water as quickly and calmly as possible.
- Do not enter the water if bleeding from an open wound, and enter with caution if menstruating.
- Avoid swimming with uneven tan lines or brightly coloured swimwear or dive gear.
- Do not harass or provoke a shark if you see one.

Take action

If you see a shark while in the water, the best thing you can do is stay calm. This may seem impossible or difficult, but it serves several

It is recommended not to swim in water where baitfish are aggregating – sharks, such as this blacktip reef shark (*Carcharhinus melanopterus*), may be actively hunting in these areas. Photo by Elizabeth Perkins.

purposes. First, overreacting and panicking may lead to erratic movement or increased heart rate, which may actually attract an otherwise uninterested shark. Most sharks are not harmful and are uninterested in (or even fearful of) humans, so will swim away in their own time. Some may come close to a person out of mere curiosity, but will not display threatening or adverse behaviour. Thus, it is best to remain calm and relatively still. If the shark is large and of a potentially dangerous species but is not in the immediate vicinity, then controlled action should be made to exit the water wherever possible. This should be done while keeping the shark in sight. If the shark can be seen, certain behaviours characteristically suggest the likelihood of an immediate attack. These include rapid rushes towards you, jerky swimming in a side-to-side, zigzag, or up-and-down motion, pectoral fins lowered from their normally near-horizontal position to as much as 60° downwards, and arching of the back (Miller and Collier 1980;

I've never actually been involved with a shark attack myself. But, having said that, as a full-time lifeguard on the beach in Queensland, I have encountered many sharks as a result of spending so much time watching the water. The most challenging aspect of working in an environment with the potential for a shark attack is constantly being alert about what could occur and knowing what to do in the situation. If there is a major attack, getting paramedics to the scene quickly is obviously a must so the patient can receive advanced care as soon as possible. People can die very quickly from blood loss if the correct care is not given to them.

People are obviously aware there are sharks in the ocean, and sharks always receive a lot of media attention. One of the most frequent questions that we're asked as lifeguards is about sharks in the area and what people need to worry about. Recent media coverage has definitely made a lot of people more cautious about being in the water. And when there's more of a media frenzy about sharks in certain areas, I can definitely see a change in where people do and don't want to swim. At times, the media can hype up reports to gain attention and traction, and sharks have the tendency to do that. The media can sometimes go a bit overboard about shark sightings, but the public should know where sharks have been active and if there has been an attack. I suppose that increasing knowledge and awareness is a good thing.

My best advice to beach users in terms of sharks is that it's important to be smart about when you should and shouldn't enter the water. If you know there have been baitfish or larger fish schooling in the area, there is obviously a chance that sharks might be around chasing food. Personally, I wouldn't surf or swim near river mouths on the outgoing tide. With the water flushing out, there seems to be a lot more fish activity building up in those areas due to the abundance of potential food flowing out of the river. This, in turn, could potentially lead to more shark activity. Being in the ocean as often as I am, you tend to see a lot of marine life. I often surf early in the morning, before the crowds pick up, and I don't tend worry too much about sharks. Having said that, I'm mindful of fish schooling in the area or the presence of dead animals (e.g. whales, turtles) and I won't enter the water during those times. I also like to make sure I'm not swimming in really dirty water near river mouths, because I know that sharks don't have great eyesight and will sometimes use their teeth as a means of testing potential food items.

There are clear local trends in both shark and baitfish activity patterns, and these have been consistent over the past 10 years. From Easter school holidays to about August you see quite a large amount of baitfish and other schooling fish coming through. I've noted different types of sharks tend to come to shore at different times of the year. Summer seems to involve a large amount of leopard sharks basking on the surface, and there appears to be more sightings of tiger sharks in the warmer months as well. We also tend to see more bronze and dusky whaler sharks in summer. In winter, we typically see more of the bigger sharks close to shore, chasing the schooling fish. There are often schools of mullet that travel out of the river system and out to sea to find the breeding areas, and we get a lot of sharks chasing them. These are mainly bull sharks, of medium to larger size. We also have the odd great white sighting when the whales are around.

Sharks are vital to the ocean ecosystem. If we were to cull sharks, there's a chance another predator could come into the shallows to feed. There are drum-lines in place off our beaches in Queensland to help protect beachgoers, which I feel is quite an effective measure. More aerial observation could also be a solution for people to become more aware of where sharks are during the day, which could help them to determine where and when to get into the water. If there's a confirmed shark sighting, lifeguards will shut the beach and advise all swimmers and surf craft to leave the water in the interests of safety. We do our best to track the shark and make certain it has left the area, using a jet ski or inflatable rescue boat to follow the creature for as long as we can see it. We also have the ability to task the Westpac Lifesaver Rescue Helicopter to fly over and scan the area for any sharks. As per our policies, a beach will remain closed for at least 30 minutes after the last confirmed sighting of a shark. I've found that people are generally quite responsive to our directions and more than happy to oblige when there's a shark in the water.

Tim Wilson
Lifeguard, Surf Life Saving Queensland

ISAF 2017). If the shark is already close and retreat is not an option, then quietly maintain your position or slowly back up against a large static object (e.g. a rock) if possible, to provide a sense of greater size and to reduce the number of angles from which the shark can approach you. If you have been fishing, and are currently holding a fish, then discard it and quietly and gently move away from the area. If you are with a buddy or companion, then maintaining a back-to-back position allows for someone to consistently face the shark, despite its movements. If you are alone, try to keep the shark in sight as much as possible. Some species of potentially dangerous sharks seemingly prefer to avoid their prey's field of view; they typically attack from behind rather than from the front (Baldridge 1988; Collier 1992; Levine 1996; Byard *et al.* 2000; Ritter and Amin 2014). In the case of bull sharks (*Carcharhinus leucas*), it has been shown that maintaining a vertical rather than horizontal body position is more effective at reducing close encounter rates and keeping sharks at a greater distance (Ritter and Amin 2012). The second benefit to staying calm around sharks is to maintain a sense of control and rational thought. The more you can think through the situation, the better your chances of being able to devise a strategy to fend off the animal if it does attack. Maintaining a sense of control, through conscious thought and action, also assists with mental recovery following an encounter.

If you are attacked by a shark, the best thing to do is to stay calm and take action. Talking to survivors shows that this seems to be a subconscious response anyway. Take any opportunity to get away from the shark and out of the water. Defensive action may work initially, but a highly motivated shark may return after an initial attack, and defensive actions may be decreasingly effective with repeated use. The shark's motivation factor will be the biggest driving force in the effectiveness of fighting back. If the animal is starving, and obtaining an energetic meal is a life-or-death situation for it, then fighting it off may be more difficult than if the shark is simply curious and investigating the scene through its use of taste and touch. Other factors, such as the species (and individual personality), size of the animal, location, and human support available, also influence the degree of success with retaliatory measures. While taking action is recommended, it may not be required, or even an option in many shark attack situations. As the majority of attacks are hit and run, the whole interaction may be over before a person even realises what has happened, making retaliation impossible and unnecessary. However, it is still advised that people get out of the water following this type of incident.

Some sources recommend repeatedly hitting an attacking shark on the snout. For some species, and some motivations, this action may be enough to deter the shark. Shark skin is covered in rows of small triangular-shaped scales (dermal denticles), which are more like teeth than fish scales, and the skin of many species is extremely tough, abrasive and highly impenetrable. However, many species also have an extensive array of sensory structures on the snout, so this area may be more sensitive than many other parts of the body. The use of any sort of weapon or equipment accessible for hitting an attacking shark is highly recommended, due to the abrasiveness of shark skin, the added strength and durability provided by an accessory object, and the lower consequence of something other than your hand being taken if a misguided or mistimed blow to the animal results in a bite. If accessible, though, it is generally considered to be far more effective to target the shark's eyes or gills rather than the snout, as these areas are much less protected (thus more vulnerable) and are more vital to the survival needs of the shark.

Personal mitigation devices

Although there are number of things that a person can do to lessen the risk of shark attack, the reality is that an attack can still occur. Thus, further protection through the use of a directed personal shark mitigation device may be beneficial, or at least may help to reduce any perceived threat of such an event, allowing individuals to more thoroughly enjoy an activity. Personal mitigation devices operate on the premise of making a person undesirable (through appearance or an external stimulus) or by effectively hiding a person from shark sensory modalities. With global shark attack incidence rising, sales in personal shark deterrents and repellents are beginning to gain traction as people look to take added action to protect themselves.

There is a wide variety of personal mitigation devices commercially available, with numerous others under development. However, few have been through independent testing to assess their effectiveness in deterring sharks under natural-use conditions. As a result, there is very limited information or support available for consumers to help guide decisions on whether a particular shark deterrent might be suitable (and effective) for their specific needs. Many of these devices are very expensive, often costing several hundred dollars.

Most personal shark attack mitigation devices are based on the manipulation of shark sensory capabilities. They promote personal protection, without the need for lethal measures. Varying degrees of success have reportedly been achieved through personal mitigation devices, but current technology is restricted by our limited understanding of shark sensory biology, the wide variation in the sensory capabilities of different species and our understanding of the factors that drive sharks to bite humans.

Electrical

Electrical repellents are designed to exploit the fact that sharks are significantly more sensitive to electrical fields than humans. Thanks to sharks' ampullae of Lorenzini, they are able to detect extremely weak electrical potentials (as low as 1 nV/cm) generated by geological or biological stimuli (Kajiura and Holland 2002; Jordan *et al.* 2011). Sharks primarily use their electrosensory system to locate and home-in on prey, and this system generally becomes highly sensitive at close proximity (within 50 cm) (Hart and Collin 2015). Due to its incredible sensitivity, the system can be artificially overwhelmed through receptor oversaturation, which is what electrical shark repellents aim to do.

The best-known (and well-trialled) shark mitigation device currently available is the Shark Shield. The Shark Shield is the latest technology from Shark Shield Pty Ltd (Adelaide, Australia; https://sharkshield.com/), replacing the SharkPOD (protective oceanic device) technology originally developed by the KwaZulu-Natal Sharks Board. There are currently three models of Shark Shields available to suit the activities of diving (and related activities), technical diving and surfing.

Shark Shields use an electrical field to deter close shark interaction. The main unit either attaches to the user's ankle and trails a 2.2 m long antenna with an elongated electrode plate at each end, or fits into the kicker of the tailpad of a surfboard and has a stick-on adhesive decal antenna. The efficacy of the device has been independently trialled through scientific studies (Huveneers *et al.* 2013; Kempster *et al.* 2016). When static baits were set at distances of greater than 2 m from the electrode, the presence of the electrical field did not deter white sharks (*Carcharodon carcharias*) from consuming the baits, although it did result in increased consumption time and fewer interactions per approach (Huveneers *et al.* 2013). However, the proportion of baits taken within 2 m of the electrical field was reduced. Further investigation supported the close-range deterrent capacity of the device, showing an 83% decrease in interactions with static baits within 2 m of an active Shark Shield, compared with baits around non-activated control devices (Kempster *et al.* 2016). The average distance that sharks got to the bait before turning away in the presence of the Shark Shield was 82 cm. The importance of this information is that, with diving models of the Shark Shield, the centre of the electrical field trails the diver by over 1 m, thus the majority of the body would still be exposed and not protected by the electrical field. This needs to be taken into consideration by manufacturers and users. A more efficient design would connect the trailing end of the antenna to a dive tank or other apparatus at waist level, to keep the entire person in the maximally protected area.

Further results of those studies reinforce the notion that the functionality of any mitigation device will be largely impacted by the animal's motivational state and the energetic cost involved in attempting to bite the prey. While bites from inquisitive sharks may be deterred to an extent, active predatory bites may still be undeterrable. Bites on relatively slow, defenceless humans would not require a great deal of energy expenditure for large sharks, and as such, even inquisitory bites may still be considered opportunistic, despite the abnormality and discomfort of the localised electrical field.

To help quell the fear of participants, Shark Shields were attached to the course buoys at the 2016 Cole Classic swim in Manly, New South Wales. However, it was later acknowledged that they would have been ineffective in shark attack mitigation unless attached to the actual swimmers. Thus, they were not reinstalled for the 2017 event. Instead, event coordinators decided to rely on patrolling personnel and lifeguards to look for and raise alerts at any sign of sharks (*pers. comm.* with event staff).

Other electrical deterrents, such as the RPELA from Surfsafe, are also on the market but have not undergone similar levels of testing.

Another variety of electrical deterrent is magnetic or electromagnetic repellents. Unlike deterrents that rely on the generation of electrical potentials, magnetic deterrents do not require a power source; instead, they utilise the electrical fields generated by electropositive metals and permanent magnets. It has been proposed that a shark's ampullary system may be overstimulated by the induced voltage of a magnet through the process of electromagnetic induction (Kalmijn 1982; O'Connell *et al.* 2011). Electropositive metals react strongly to sea water, producing a small electrical current. Research has predominantly focused on these resources for use as shark deterrents in commercial longline fisheries (which traditionally hook an overwhelming amount of shark bycatch). However, these metals have provided inconsistent results in trials and between species. In some trials, such as when bait was affixed to the naturally electrogenic lanthanide metal neodymium, sharks showed clear confusion, which took the form of repeated bites and a longer time spent trying to remove the bait (McCutcheon and Kajiura 2013). Obviously this would not suggest an effective shark attack mitigation measure. Furthermore, electropositive metals produce only a short-range electrical field (<85 cm; Hart and Collin 2015), which may not provide enough distance to deter a rapidly moving, motivated animal. They also corrode quickly in sea water, leading to the requirement for regular replacement.

Sharkbanz is a personal deterrent that incorporates magnetic technology into a wrist or ankle band. The device can be worn while swimming, diving, surfing or doing other water-based activities. It does not require batteries, electricity or chemicals. However, the effective range of the product is likely quite limited. While these bands are marketed as effectively deterring hit and run attacks, particularly those occurring in murky waters where the visual sense would be limited, it seems unlikely that a large, motivated shark would be inclined (or able) to abort a rapid ambush attack at the very limited distance at which these magnets would cause functional discomfort. However, the technology could be beneficial in deterring smaller, slower reef sharks, such as those commonly encountered in the coastal waters off Florida, from swimmers and waders. Of note, though, a victim of a recent shark bite from what was likely a small reef shark was wearing a new Sharkbanz bracelet when bitten in the Florida surf (Robbins 2016).

Electropositive metals for personal deterrents have been discussed, but the results of these metals as deterrents are mixed and their effective range would be 85 cm or less (McCutcheon and Kajiura 2013).

Chemical

The first funded shark attack mitigation testing began during WWII in the early 1940s, when the US military sanctioned the development of shark repellents for servicemen adrift in the ocean (Baldridge 1990; Stroud *et al.* 2014; Hart and Collin 2015). One of the first products developed was the Shark Chaser, which used cupric acetate as the active ingredient. The Shark Chaser proved ineffective as a shark repellent but may have psychologically aided the servicemen who were fitted with it (Baldridge 1990; Smith 1991). Various other chemical agents including fish poisons, chlorine, metallic poisons, irritants, ink and unspecified poison gas generators and chemical stenches were also tested, with similarly unsuccessful results (Springer 1955). While these did not deter the sharks from taking baits, in many instances they did subsequently kill them (Hart and Collin 2015). Some of the more successful early trials employed compounds such as ammonia acetate, copper salts and, particularly, copper acetate, a modified nigrosine-type dye (which was found to act as both a chemical and visual deterrent) (Fogelberg 1944; Hodgson and Mathewson 1978). However, further testing found that all these products, even in combination, were ineffective at deterring some species of sharks.

The challenges facing the development of chemical repellents primarily revolve around the rapid dispersion (and, consequently, reduced efficacy) of chemicals in vast volumes of continuously circulating water. Other significant challenges include the necessity to isolate compounds that would not only be effective at deterring sharks in very low concentrations but that are also non-toxic to the environment and other aquatic life. Effective products must be able to be stored without denaturing for a prescribed amount of time to make them of use.

Following the failures of the initial products tested in the 1940s, various other products were trialled. Although some met certain requirements for personal shark attack mitigation, none met all the relevant requirements. Over 100 chemicals were tested in the late 1950s and 1960s, including metallic salts, amino acids, nicotine, human sweat and urine – all proved ineffective (Hodgson and Mathewson 1978). Pardaxin, mosesins and pavoninins, which are naturally occurring toxins derived from soles (*Pardachirus* spp.), held promise as deterrents, but were found to lose potency during protracted freeze-drying (Hart and Collin 2015). Sodium dodecyl sulphate was also examined, but the effective volume required was too great and the broader environmental impacts were not fully understood (Baldridge 1990). Sodium lauryl sulphate (found in many household products, like shampoo and laundry detergent) has been shown to elicit a response in sharks when squirted directly into the mouth, but was proven to be ineffective when released into the water in low concentrations (Smith 1991; Sisneros and Nelson 2001).

Currently, biologically relevant compounds (semiochemicals) are the focus of many chemical deterrents (Hart and Collin 2015). Repel Sharks (http://repelsharks.com/), which is available in a pressurised aerosol can, is marketed as a personal deterrent capable of covering a broad area for a short period of time. The product is based on the long-standing anecdote that natural chemicals found in decomposing shark tissue deters sharks. Research was initiated to determine the specific chemical signals, dubbed 'necromones', that activated such avoidance behaviour. Necromones contain high concentrations of acetic acid as well as a large array of amino acids, short-chain acids, fatty carboxylic acids, amines and short-chain lipid oxidation products. Field testing of 150 mL of necromones released as an aerosol on Caribbean reef sharks (*Carcharhinus perezi*) and blacknose sharks (*Carcharhinus acronotus*) resulted in a clear cessation of feeding behaviour and temporary evacuation of an area containing a feeding stimulus within one minute of exposure (Stroud *et al.* 2014). Habituation to the stimulus was not observed, nor were responses to the bubbles, sound or solvents simultaneously produced during the release of the necromones. While effective on some species for short durations, other species that scavenge for food (which can include potentially dangerous species such as tiger sharks, *Galeocerdo cuvier*, and white sharks) may prove to be attracted to, or unaffected by, such a stimulus.

Visual

The shark visual system can operate at distances of up to 100 m; however, its functionality is largely dependent on light, environmental conditions and species-specific adaptations. Although sharks have been behaviourally shown to aptly respond to brightness contrast and differences in light intensity, it is likely that they possess monochromatic vision and are anatomically unable to support true colour vision (Hart *et al.* 2011). Essentially, sharks are colour-blind. This negates the long-time concept of sharks being attracted to 'yum yum yellow' solely on the premise of colour, although the contrast and intensity of a yellow object among blue water or sky could still be highly distinguishable. In fact, testing of sea survival equipment found that sharks (including bull sharks, tiger sharks and sandbar sharks, *Carcharhinus plumbeus*) tended to ignore objects with low reflectivity, and blue sharks (*Prionace glauca*) and shortfin mako sharks (*Isurus oxyrinchus*) readily attacked dummies in standard bright yellow lifejackets while ignoring those dressed in black lifejackets (McFadden and Johnson 1978). Before personal safety equipment is altered on the premise of shark attack, though, the question of design and purpose must be considered – are the chances of shark attack high enough to outweigh the high visibility of these colours, which is vital during search and rescue operations? I would say no. In terms of everyday swimming, diving and surfing equipment, though, the answer would be very different.

One of the more interesting personal shark attack mitigation products developed to date was the technologically simple Shark Screen, developed by a US naval scientist. Shark Screens were essentially large plastic bags (of varying colours, materials and opacity) with inflatable collars for buoyancy (Tester *et al.* 1968). The bags were designed to be flooded with sea water, and a person adrift in the ocean could climb inside. The bags would not provide complete physical protection from sharks (indeed, early trials showed that the bags were easily punctured or suffered from ripped seams that quickly led to deflation) – they were designed to visually conceal the occupant and contain any chemical stimulants they may produce, such as blood or urine, minimise body movements and isolate bioelectrical signals.

Difficulties that arose during testing made it hard to gain a true understanding of the efficacy of the bags in an actual drift situation but, during controlled tests, sharks made very few bites, bumps or rubs on the bag itself. The bags were predominantly tested with moderately sized reef sharks, and it was acknowledged that larger species may not have been so easily deterred by the presence of the bag. Sharks approached bags that were brightly coloured (highly reflective) more than those that were dark in colour (less reflective). The conclusions drawn from this study were that the protective bags provided superior protection for servicemen compared to Shark Chaser and other chemical and electrical deterrents of the time.

More recently, various wetsuit and surfboard designers have tried to exploit scientific advances in our understanding of sharks' visual acuity (the ability to distinguish separate lines at a given distance, tested in humans by reading the letter chart at an optometrist) and surface reflectance spectra (e.g. the contrast between the human and the environment) to make users appear unpalatable or largely indistinguishable from the background. Although such products are on the market, they are yet to be comprehensively tested in independent field trials. The first product along these lines was a wetsuit designed with black and white bands, intended to mimic the colouration and pattern of venomous sea snakes. The results from field trials of this design on reef sharks were mixed (Hart and Collin 2015). The same design resurfaced in the Shark Attack Mitigation Systems (SAMS) Technology™ 'Warning' range (http://www.sharkmitigation.com). The pattern is marketed as providing an overt visual message that the wearer (or user, in the case of a patterned surfboard) is not a typical food item, and may be unpalatable or dangerous for consumption. It may perhaps work on some species and in some locations (e.g. where sea snakes are endemic) but the logic behind the Warning black and white banding pattern for tiger sharks, in particular, is flawed – sea snakes are a preferred food item of those sharks (Heithaus *et al.* 2002). Furthermore, for the unpalatability/dangerous prey concept to be effective, the danger to the consumer must be relevant. Sea snakes generally utilise venom as a predatory advantage for capturing prey, not as a protective deterrent towards predators. And sea snakes are venomous not poisonous, so this adaptation would not make them unpalatable to formidable predators.

The most novel, and scientifically backed, visual deterrent to date is the other SAMS design, Cryptic. This patterning was based on the calculated reflectance spectra of studied shark species, yielding colours and patterns that would minimise contrast against the user's water background. While it is an interesting and research-based strategy, it would disrupt only the visual sensory modality of a shark – and vision is not the only sensory modality that is relied on during predation. In the case of bull sharks, especially, which regularly inhabit murky environments with low visibility, vision would not be expected to be a primary sensory modality for locating and capturing food. For maximum benefit, the reflectance spectra and patterns in the design would have to match those of the environmental conditions during use (to provide minimal contrast) in line with the visual capabilities of the potentially dangerous species of sharks in the region. Trials by independent researchers are warranted to determine the functionality of this technology on active, motivated sharks in common user scenarios in the natural environment.

Auditory

Although less common, auditory deterrents have been trialled and one product is currently on the market. Experiments in the late 1970s found that lemon sharks (*Negaprion brevirostris*) and silky sharks (*Carcharhinus falciformis*) would shift their behaviour and withdraw from a situation (defined as a 180° turn and departure) following certain auditory cues. The research was originally designed to test whether killer whale (*Orcinus orca*) screams would deter sharks, as the calls from natural predators have been shown to lead to withdrawal behaviour in other marine species (Klimley and Myrberg 1979). It was found that, while whale screams did elicit a withdrawal behaviour in some trials, a more general noise band of 500–4000 Hz at 18 dB above broadband ambient was found to be the most effective deterrent, causing shark withdrawal behaviour in 16 out of 20 playbacks.

Interestingly, and notably, all auditory cues failed to elicit behavioural responses in water temperatures below 21°C.

It is vital to note that low-frequency, broad-bandwidth, irregularly pulsed sounds can have the opposite effect and may actually attract sharks (Klimley and Myrberg 1979). The increased boat traffic following the construction of Suape Port in Brazil is a significant example of this effect in a practical situation. The novel auditory cue, along with related environmental changes, are considered to be major factors in the commencement of negative shark–human interactions in Recife in the early 1990s (Hazin *et al.* 2008).

The only current commercially developed product that utilises auditory cues is the SharkStopper® (http://www.SharkStopper.com/). While predominantly designed on a small scale for personal or watercraft use, SharkStoppers® are also being developed on a larger scale with the aim of using them to deter sharks from fishing nets and gear. The personal SharkStopper® emits pulsed sounds at a frequency of 30–500 Hz or 200–1500 Hz (Hart and Collin 2015). According to the product's website, the device has successfully repelled a variety of species, including white, bull and tiger sharks. However, as with most of the other products, no independent field test results have been published. As sharks have been shown to habituate to sounds (Myrberg *et al.* 1969, 1978), rigorous field testing would be necessary to show how long an auditory repellent would be effective for, and whether that period would be sufficient to allow a person to safely carry out their intended activity, or at the very least evacuate the water before a shark returned following an initial encounter.

Review of personal mitigation devices

A variety of personal mitigation devices were recently reviewed by the Australian independent product testing company, Choice (https://www.choice.com.au/health-and-body/diet-and-fitness/surfing-and-snowboarding/articles/shark-repellents-review). Choice did not test these products around sharks; the results are based on reviews and product information. A summary of the findings is presented in Table 6.1.

Table 6.1. Comparison of personal shark mitigation products, tested by Choice (https://www.choice.com.au/health-and-body/diet-and-fitness/surfing-and-snowboarding/articles/shark-repellents-review).

Category	Product	Cost	Use	Fit	Review
Electrical	Shark Shield	AU$749	Diving, surfing	2 m cord that wraps around the ankle, or is installed onto the board/kayak	Found to be mostly effective during independent testing on white sharks. Anecdotal evidence of shark bites, despite product use.
Electrical	NoShark	US$399	Diving, swimming, surfing, snorkelling, other	Anklet	Unlikely to emit a pulse strong enough to effectively deter sharks due to the close proximity of the electrodes.
Electrical	Surf Safe	US$389	Surfing	Electrodes installed onto the surfboard	Awaiting independent testing.
Magnetic	Sharkbanz	AU$149	Diving, swimming, surfing, snorkelling, other	Wrist or ankle band	Reported to repel a variety of sharks, but not recommended for white sharks or large tiger sharks. Unlikely to be effective according to independent scientists, and inconclusive as a repellent in the commercial fishing industry.
Magnetic	Shark Shocker	US$35	Diving, swimming, surfing, snorkelling, other	Wrist or ankle band	Unknown. Unlikely to be effective according to independent scientists, and inconclusive as a repellent in the commercial fishing industry.

Category	Product	Cost	Use	Fit	Review
Acoustic	SharkStopper®	Not yet released	Diving, swimming, surfing, snorkelling	Small plastic leg band	Produces a 'multi-patented acoustic sound' mixing orca calls and a special frequency. Possible limitations, including the use of other sensory modalities to override the fact that a human is not a whale, and the concern that this theory would not be effective over a range of locations in the wild.
Chemical	Anti-Shark 100 (SharkTec)	US$25	Diving, swimming, snorkelling	Aerosol can	Requires a pocket to carry. Acts to create a temporary safe zone, until the product disperses. Only functional in the event that there is an opportunity to spray it before an attack.
Visual	Dark-coloured surfboards	Various	Surfing	Board colour	Surfboard colour would not change the silhouette of a diver when viewed from directly below, so of limited potential.
Visual	Shark Attack Mitigation Systems wetsuits	AU$495	Surfing, diving, snorkelling	Wetsuit pattern/colour	Patterns based on camouflage and dangerous food options/symbiotic species. Awaiting independent testing results.

Overall comments: No products are marketed to be 100% effective. Unlike many other safety products, there are no standards for shark attack mitigation devices to guarantee that they have been adequately tested for efficacy and safety, and very little independent testing has been done on these products. The use of shark-repelling devices should not be interpreted by the user as validation for employing more risky behaviour. A personal shark deterrent may provide a person with peace of mind, allowing them to more readily enjoy a water-based activity, and in that sense is worth the cost. However, that is just about the only guaranteed benefit (if personally applicable), and other free measures (such as when and where you swim) may be just as effective in reducing your chance of being bitten by a shark.

Public perception of personal shark attack mitigation

Not surprisingly, the views expressed by the general public in response to a poll on personal shark attack mitigation were mixed (Gibbs and Warren 2015). Some felt that such precautions were sensible and rational, whereas others felt regret at the need to change their practices to accommodate increasing risk from sharks.

Following a spate of shark attack fatalities in Western Australia in 2011 and 2012 (and the associated media coverage), 43% of survey respondents indicated that the events had changed their ocean use; the remaining 57% said they had not (Gibbs and Warren 2015). The majority of ocean users were willing to alter their practices to reduce risk and felt that maintaining awareness and acting to mitigate the risk were simply a requisite actions when using the ocean. In an attempt to reduce the likelihood of encountering sharks, respondents most commonly reported utilising the strategies of conducting their ocean-based activities in groups/with others (57%) and avoiding recreational ocean use late in the evening (33%). Other responses included remaining close to shore, avoiding deep water and river mouths, avoiding particular locations (e.g. near marine mammal colonies or baitfish) or times (e.g. during whale migrations, after heavy rainfall), not using the water in overcast conditions or at dawn or dusk, monitoring various shark warning systems and avoiding areas where sharks were reported, wearing shark repellent devices, only using patrolled beaches and reducing the frequency of ocean usage. All these strategies would be beneficial in mitigating the risk of negative interaction with sharks, and demonstrate the ability and initiative of individual water users to simply make well-informed choices to lessen their personal risk.

Conclusion

The unfortunate truth is that, similar to all animals, sharks are unpredictable and unmanageable. Therefore, the risk of being bitten by a shark when in the water will always be present. However, there are a number of ways to reduce the likelihood of negative interaction. The most practical way is to simply make smart choices when carrying out recreational activities in areas that overlap with sharks. While no personal mitigation device has proven to be 100% effective, some have shown the potential to be beneficial in certain circumstances. Possibly the most important function of these devices, however, would be the added peace of mind that they may provide. With risk being so minimal in the first place, if these devices allow a person to enjoy their normal recreational activities without a constant fear of sharks, then this is an advantage. Knowledge of the local environment and the shark species that could be encountered (in addition to the user's intended activity) is imperative in selecting the most appropriate personal mitigation device. It is likely that, as these devices become more in demand, and as our knowledge and understanding of shark biology, movement and motivation continue to grow, they will become more effective. To encourage use, appropriate mitigation devices should be available for hire along with recreational equipment (e.g. surfboards, scuba equipment, non-motorised watercrafts).

7

Shark bite first aid and trauma medicine

Shark attack first aid

Although common public perception suggests that a high proportion of shark bites are immediately fatal, this could not be more unrealistic. In reality, very few bites prove fatal, with only four deaths worldwide in 2016. However, the likelihood of survival following major shark bite trauma is pivotal on the victim promptly getting to shore, having bleeding from major surface injuries controlled and having their plasma volume expanded. The decreased fatality rate of shark bite victims over the years is highly reflective of the improved training and preparedness of first responders in attending to these sorts of injuries. Although the extent of shark bites can vary immensely, it is highly beneficial for shark attack victims with significant injuries to be treated at the scene; in more traumatic cases (which this chapter mostly focuses on), survival hinges on appropriate pre-hospital care (Goodwin and White 1977; Ricci et al. 2016). Although injuries may be similar to more common sources of trauma, such as motor vehicle accidents, the immediate condition and environment of shark attack victims differs. Following a major shark attack, the person is often in a state of extreme exhaustion from fighting the animal or from making their way out of the water with massive blood loss, debilitating injuries or possibly even limb loss impairing function. And, having been in the water, they are often hypothermic and may have been nearly drowned during the attack.

If you are at the scene when someone is bitten by a shark, or are the first responder, there are various things you can do to assist. Being at the scene of a major shark attack is highly traumatic for bystanders and first responders, but it is vital that everyone present remains as calm as possible, and acts quickly and smartly. First aid should proceed rapidly, following the principles of trauma assessment and treatment. The patient should be reassured continuously. Depending on the extent of the injuries and the condition of the patient, they may or may not be able to walk or move themselves. Thus, the first thing that may need to be done is to remove or assist them from the water. As time is of the absolute essence, and to prevent any excess muscle movement, the victim should be moved only as far out of the water as is necessary to commence immediate resuscitation (Caldicott *et al.* 2001) and to have better control over the working environment (e.g. not in a position where an incoming tide may reach in 10 minutes). Dragging open injuries through the sand should be avoided or limited as much as possible – placing the victim on a blanket or towel may be useful to help achieve this. The highest priority, as with any trauma, is to secure the airway, followed immediately by controlling any bleeding. This can be vital and life-saving in major traumas. The best and most preferred approach for controlling bleeding is to elevate the limb and apply direct pressure above or directly on any bleeding points. While tourniquets are not generally preferred, they are often the most appropriate option for large open injuries or missing limbs (W. Butcher, *pers. comm.*). Tourniquets may be improvised with whatever can be found at the scene, such as surfboard/body-board leg ropes, clothing, towels or shoelaces. One positive aspect of shark attack traumas, compared to many other forms of major traumas, is that external bleeding can often be easily controlled by compression or tourniquet and there is little danger from unseen internal bleeding, lung injuries that could compromise gas exchange and oxygen transfer through the body or compounding head or neck injuries (Goodwin and White 1977). Once bleeding has been attended to, emergency responders (e.g. paramedics) should be called. Of course, if there is more than one person available to help, someone should call emergency responders immediately while other people perform first aid on the patient.

During resuscitation, the patient should be protected from the elements and the effects of hypothermia as much as possible. This could mean lying them on a blanket, towel or surfboard bag, and covering them to control heat loss and to protect them from the sun, wind and sand. With major traumas resulting in excessive blood loss, the patient should ideally not be moved from the beach until circulation has been stabilised, to prevent circulatory collapse and hypovolaemic shock (Goodwin and White 1977). Hypovolaemic shock (hemorrhagic shock) is life-threatening and has been shown to be the primary cause of death following shark bite (Woolgar *et al.* 2001). It can occur when a person loses more than 20% of their blood or fluid supply, making it impossible for the heart to pump a sufficient amount of blood throughout the body. However, in more remote or isolated locations or if there is likely to be any sort of delay from medical responders, the patient should be transported by first responders either directly to a hospital or to a meeting point arranged with paramedics as the timely administration of fluids and blood is essential for survival. Ultimately, major traumas need hospital care, and the quicker a patient can get to advanced treatment the better their chance of survival. Receiving hospitals should be given as much notification as possible about the impending arrival of a shark attack patient. Depending on the region, the extent of the injuries and the hospital's preparedness for this sort of trauma, the hospital may even send its own physician or a specialised retrieval team to the scene.

The arrival of advanced first aiders (e.g. lifesavers) and paramedics allows for more advanced pre-hospital care, including the equipment, expertise and capacity to intravenously administer fluids (and in some places, the option for blood administration), oxygen, and more sterile or purpose-designed first aid materials, plus other monitoring and medical equipment and expertise. At this time (assuming the bleeding has been sufficiently controlled), the administration of fluids for volume resuscitation is the highest priority.

'Shark attack packs' have been used at surf lifesaving clubs in South Africa (Goodwin and White 1977), and have been proposed for some beaches in New South Wales (Australia). The packs contain all the equipment necessary for first aid on shark bite injuries and for administering fluid resuscitation immediately, on-site, by trained responders. In South Africa in the 1970s, arrangements were made

I work as a Community Paramedic in Esperance, Western Australia. The first patient I attended to was apparently attacked by two sharks at once. Parts of both of his arms were amputated, and both legs were mauled. By the time I arrived, bystanders had given first aid, applied tourniquets, put the patient in the back of a station wagon and driven him about 10 km along the beach, where I met them. Everyone on the scene was distressed and very anxious, but it was not to the point where they couldn't help or were paralysed. The beachgoers and rescuers who had taken immediate action with the limited resources they had really did a great job in getting the treatment process started. These were quick-acting, resourceful people who ultimately saved the patient's life.

Personally, I felt an immense amount of pressure. This patient had survived the last 20 minutes, or so, against difficult odds. He had parts of both arms ripped off. His legs were shredded, and he was barely conscious. I was the only medical person there initially, and I had to make sure that he got to the hospital alive. Although I have been a paramedic for more than 20 years, a shark attack is something quite different to the average job. There is almost a surreal element to it. It just has that aura about it. You know it's not going to be a simple band-aid fix; it will either be fatal or really bad. You also know that you're going to be scrutinised for everything that you do because of all the media attention.

Every situation we attend is different and every person is different. We go to car accidents regularly, but shark attacks are just so rare. You certainly never wake up expecting to be called out to a shark attack. The challenge was definitely present with this case. If it had just been one limb, it wouldn't have been as bad, and a leg is not so bad. But in this case, it was both arms. There were multiple sites that could have been life-threatening. The response is that much greater, to make sure you've attended to everything; you don't want to miss anything. I have treated other patients with amputations, often from industrial, motor vehicle or train accidents. But this one was a guy on a beach with his torn wetsuit still on him. He was covered in sand, and I was told how two sharks had just ripped him to bits. It's different. It's hard to put it succinctly in writing, but it's different. But ultimately, there was not really any time to think too much about what had just happened to him, I just got to work. You get in and do what needs to be done. It's not until later on when you have time to think it through and process things. I couldn't help but feel for this poor guy. He was just having a surf at a beautiful beach. He was young. Then, in an instant, his world was changed forever, for the worse. I see a lot of self-inflicted injuries from people being drunk, getting into fights or making stupid decisions. But this guy surfed that morning, a normally perfectly harmless activity, and this happened to him.

The main hallmark of shark attack patients, from my limited experience, is external bleeding. This is compared to other forms of trauma that often include blunt trauma, resulting in internal bleeding. External bleeding is simply stopped by direct pressure, covering with a dressing and elevation. The patient then needs to be treated for hypovolaemic shock (loss of blood volume). The other distinctive trait of shark attack trauma is that the patients are usually cold. They have been in the ocean, which is a cold place. Then they lose blood volume, which is the agent that transports warmth around the body through warm blood, so they can get even colder very quickly. They are also sandy. They often get dragged up the beach, out of the water, and collect sand all over them. This sometimes adds to the difficulty of treating them. At the scene, for instance, you need to keep the sand out of IVs and your equipment. You need to take extra effort to clean the site of sand before you inject the patient. These are just further complications that can delay life-saving treatment. Also, because the patient is in the water for a period of time, there is a risk of extended blood-clotting time. From my experience, human blood does not clot very well in water, or the process is at least greatly slowed. So while the patient remains in the water, blood continues to pour out without natural clotting taking place. My patient was several hundred metres from shore when he was attacked. He had to swim, without the use of either of his arms, for a considerable distance. All the while, he would have left a trail of blood because there would have been no blood-clotting taking place.

When the ambulance arrived, I went in the back with him, and continued to manage his situation during the 10 minute trip to the hospital. This included pain relief, attempts to give intravenous fluids, oxygen, looking at wounds to make sure the bleeding had stopped and checking above the tourniquets to make sure there were no further blood loss issues. I was also continuously assessing his conscious state and reassuring him. Upon arrival at the emergency department, he was given fluids and then whole blood, had his wounds dressed and was prepared for further transport. His transport included another 20 minute ambulance ride to the airport with an additional doctor and nurse, a handover to the Royal Flying Doctor Service, and then a 1.5 hour flight to Janda-kot Airport, to then be transferred by another ambulance to the Royal Perth Hospital.

The biggest mental challenge ended up being dealing with the media frenzy afterwards. Shark attacks attract their own PR frenzy, which I found out over the next couple of days. I was asked to attend live media interviews that were being broadcast nationally. Everyone was ringing, trying to get the inside scoop. It was impossible to carry out my normal job for the first 48 hours after the attack. I'm used to doing my job under all sorts of difficulties and conditions and outcomes, but this was almost like I had to change roles and become a media presenter.

Just out of interest, I checked the shark tracking website on the year anniversary of the attack, and ironically, there were two tagged white sharks at exactly the same place as where he had been attacked.

Another shark attack

I had hoped I would never be in that situation again. But earlier this week, two and a half years later, I found myself speeding to the same site to attend to yet another shark attack. But this one was different. When I got to the beach, I saw a young girl's body surrounded by several people frantically performing CPR. The 17-year-old girl had been dragged from the sea, pulseless, with her left leg missing. A tourniquet from a surfboard leg rope had been quickly fashioned. Bystanders and family members had been doing CPR for around 30 minutes.

I applied the cardiac monitor pads to see if the patient was in a shockable rhythm, but there was nothing to shock (asystole). I gave the person doing the breaths a bag valve mask and attached it to oxygen. She indicated she was a nurse and fortunately knew how to use it. I checked the amputation and tourniquet quickly to make sure no blood was escaping. There was not a drop to be seen anywhere. I then set about to get IV access. Her limbs were unusable, as there seemed to be no blood in her arms and there was no sign of any veins at all. I attempted her jugular vein in the side of her neck. I got one of the police officers to hold her hand at the base of the neck to try and help trap what little blood was in the vein so it would swell and I could insert a cannula, but this also failed. There was just no blood left anywhere, and all of her veins were collapsed.

I next went for my interosseous drill to allow me to drill directly into the bone to get fluids and drugs into her. I successfully drilled straight into the tibia. I got the other police officer to squeeze a 1000 mL bag of fluid into the patient, as I started to give adrenaline to try and kick-start her heart back into action.

By this stage the 4WD ambulance had arrived. I made the decision to continue CPR to the hospital, which was another 20 minutes away. The patient was cold and needed to be reheated before all efforts could be exhausted. I inserted an advanced airway to make ventilating the patient easier and continued to inject adrenaline every four minutes, willing her to come back to life.

Everything was tried at the hospital, but nothing could bring her back. I had known this the moment I saw her, especially given 30 minutes of CPR had already been performed. But I refused to think about this, knowing we still had to try, just in case she could miraculously pull through.

After leaving the hospital, I still had to complete hours of cleaning and restocking equipment, filling out paperwork and debriefing the ambulance crews. The wet sand had gotten into everything. I finished just after 10pm – six hours after receiving the initial call. I then rang the police officer in charge and discussed the media interview planned for the morning.

The next day was déjà vu. The media conference had lots of TV cameras with journalists from Perth that I recognised from the previous shark attack. But this was different. Last time the person had escaped with their life, even though he had lost parts of both arms. This time, a 17-year-old girl was dead.

A couple of days after the tragedy I went for a walk along the town's main surfing beach, West Beach. There were several surfers doing what they loved, surfing some waves. You will never keep people out of the beautiful turquoise water along the southern coast. I still live in hope that I won't have to attend anything like this again, but I know it's a real possibility.

For anyone who does find themselves in the situation of being a first responder to a shark attack victim, the most important thing is to stop any bleeding. That might mean applying a tourniquet. And this might mean improvising to do so. The next step is to get an ambulance to the person. In rural and remote locations, this may mean transporting the patient yourself for at least some of the distance. Sometimes people are reluctant to move the patient, but if bleeding is the problem, time can make the difference between life and death. The 'Emergency+' app is also important. It is a free app for smartphones that provides a person's exact latitude and longitude so any emergency service in Australia can send a response to the precise location.

I am completely in favour of greater mitigation measures being carried out, and particularly culls of species like great whites and tiger sharks. Culling is the only method that truly works, as it removes the threat completely. I believe there would only be good to come out of there being no more man-eaters left. Maybe this is very simplistic, but when a big shark attacks a human, it leaves a major impression. Humans before sharks, is my belief.

Paul Gaughan
Esperance Community Paramedic, St John Ambulance

between the Natal Blood Transfusion Service and the Surf-Lifesaving Association of South Africa for human albumin solution to be kept available at beaches. Normally, due to the short shelf-life of the solution and the rare occurrence of shark attacks and other major traumas on beaches, this would result in unjustifiable wastage of such a valuable and expensive resource. In recognition of this, the lifesaving association continuously rotated stocks back to the transfusion service before the expiration date, so that there was no wastage.

Regardless of the extent of a shark bite injury or the first aid received, it is imperative that anyone bitten by a shark is seen by a doctor as soon as possible for further medical assessment and treatment. While nearly half of all shark bites are minor and result only in punctures or lacerations to the skin or soft tissue, antibiotic therapy, washing out or debriding the site(s) may still be required (Lentz *et al.* 2010).

Advanced treatment for shark attack injuries

Medically speaking, shark attack injuries are just another scratch, abrasion, trauma, exsanguination or amputation, and the treatment of these injuries follows normal medical protocol appropriate for the extent of the injury. Indeed, the severity of injuries sustained in major shark attack traumas are often compared with high-speed motor vehicle accident traumas, blast traumas and military injuries. However, some unique characteristics directly related to shark bites have been found, contributing to our overall understanding of the most important principles in the medical management of shark attack. In addition to the need for early resuscitation, the use of presumptive and continued antibiotic therapy, rapid transport to a point of definitive care, and aggressive surgical intervention including complete tissue coverage are all essential for optimal results (Caldicott *et al.* 2001; Ricci *et al.* 2016).

The two most distinct and challenging components of shark bites for treating clinicians are the extensive amount of tissue damage, which is frequently associated with hypovolaemic shock, and the excessively high incidence of atypical wound infection.

Shark bite injuries

Two scoring systems have been developed for comparing and communicating the extent of shark bite injuries. The first is the Durban classification system (Davies and Campbell 1962; Table 7.1), which ranks the grade of injury on a scale of I to III, based primarily on the number of isolated body locations injured and the location of those injuries. A score of I generally suggests a fatal prognosis, whereas a score of III suggests survival, assuming appropriate treatment is received. The second system is known as the Shark Induced Trauma (SIT) scale (Lentz *et al.* 2010; Table 7.2). It ranks shark-induced trauma on a scale of 1 (minor) to 5 (likely fatal), based on clinical finding criteria such as blood pressure, location and depth of injury, extremity or organ loss of function, treatment and fatality. Other categorisations of shark bite injuries often focus on the number and extent (e.g. size/depth of areas of missing tissues, bruising, abrasions, lacerations) of injuries and the bodily location of the injuries (e.g. head, neck, torso, arms, legs) (Ricci *et al.* 2016).

All trauma can be horrific, but I guess the very thought of being eaten adds a macabre element to the scenario, and I think we can all relate to the horror of being attacked. The thought of being prey is very different from other forms of trauma, it's very primal. And since seeing the reality of a shark attack firsthand, I must admit that my fear of shark attack has amplified. The actual wounds can be very graphic, and they vividly display the immense power and ferocity of an attack by these creatures. I live on the Gold Coast (Queensland, Australia) and have swum, snorkelled, surfed and dived in these waters from an early age. I used to find sea snakes to be more threatening than sharks when diving, but now, having seen extensive shark-related injuries, I have since changed my mind. The destruction of human tissue by razor-sharp teeth, often in the classic 'half-moon crescent' shape is instantly recognisable.

I have been fascinated by sharks from an early age and I remember avidly reading stories from people like Rodney Fox. I don't really feel any different about sharks now than what I did before attending to shark attack patients, but I guess that I am now more personally aware that they are out there and of the power they possess. Sharks are beautiful animals, and very well adapted to their environment. They are first-line predators. If you enter their domain, you may become their prey. Most patients are very accepting of this and wish no harm to sharks. I, myself, disagree with reactive hunting or culling of sharks when an attack has occurred. If you want to enter the water and not be at risk of shark attack, then swim in a pool; but you won't catch any good waves or see any spectacular sea life that way. I just returned from holiday in Bali and was scuba diving while there. I did think about our recent cases before getting in the water, but found that once I was underwater, I was just so fully absorbed by the beauty of the underwater environment that I didn't worry about sharks.

Treating shark attack patients has elicited some different emotions compared to other traumatic injuries to a degree, but the medical procedures of treatment are not really different to comparable traumas. In the initial phase of treating shark bite injuries, haemorrhage (blood loss) is always a major issue and, in the case of limb injuries, it is usually the cause of death. Haemorrhage leads to a cascade of physiological threats to life that can be very complex to manage. Later, the complexity of treatment diversifies. Patients often have multiple injuries that involve more than one part or system of the body and may require the attention of many multidisciplinary health care professionals and teams.

Any severe trauma has a massive impact on the individual. It is not only physically painful, but also mentally challenging. It radically alters a person's lifestyle in an instant. Most shark-related injured patients are young and have been out there leading active, healthy lives. Next thing, they wake up in the intensive care unit with the possibility of a lifelong disability. The shark bite patients that I have attended to have all reacted in different ways. Each person has their own unique experiences and backgrounds, and this alters their perceptions. Some are quite withdrawn and reluctant to talk about the attack, while others will relay the account in graphic detail to all who ask (and sometimes you don't even have to ask). Others are more quietly accepting of the situation. But all readily agree they are very lucky to be alive! The extent of the injuries and the impacts these injuries have on the individual and their family and friends is a powerful determining factor on how they will react. I don't like to use the word 'victim' in dealing with trauma patients. I find those who identify strongly with this word are at greater risk of poorer outcomes and longer rehabilitation.

None of the staff here really had any difficulty coming to terms with the shark attack injuries that we saw. We deal with all sorts of trauma, and have done for many years, so we are a battle-hardened group. We do get to know our patients and their families and friends very well, as we case-manage them from admission to discharge, and beyond. Sometimes it's the difficulties that the loved ones go through that can pose the greatest challenge.

I feel personal protective devices, if proven to be effective, are the way to go to reduce shark attacks. I have to buy a helmet and protective gear to ride my bike, so I think that the added cost of a personal protective device against sharks would be acceptable to most surfers and divers. They could also be incorporated in life-preserving aids for personal and commercial use, such as airlines and ocean liners. I hope electronic personal protection devices are proven effective and made commercially available at a reasonable cost. Then the onus is upon the water user.

Matthew Scott
Clinical Nurse Case Manager, Trauma Service, Gold Coast University Hospital

In a large percentage of shark bites, injuries do not actually equate to a significant amount of missing tissue. The severity of shark bite wounds often depends on the size, species (relating to species-specific tooth morphology and serration) and motivation of the shark, as well as the activity of the person when bitten. Even bumps from a shark can cause significant bruising and abrasion due to the roughness of shark skin and the speed at which they travel. Such bumps can result in long, deep, even grooves in the person's skin and underlying tissue. The (assumed) erroneous bites of hit and run attacks tend to result in single injuries, often on an extremity, whereas in more serious bump and bite or sneak attacks, where the bite generally represents purposeful effort, significant tissue loss and amputation are more common. Sharks may bare their teeth and slash as a defensive mechanism, which results in linear lacerations or incisions (as opposed to the more characteristic

Table 7.1. Durban classification system of prognosis following shark attack, by wound type

Grade of injury	Wound description	Outcome
I	Both femoral arteries; or one femoral artery and one posterior tibial; or one femoral artery in upper 1/3	Often fatal
II	One femoral artery in lower 2/3; one brachial artery; two posterior tibials; abdominal wounds with bowel involvement (major)	Should survive, with appropriate pre-hospital treatment
III	One posterior tibial; superficial limb wounds; no vessels cut; superficial abdominal wounds; with no peritoneal involvement; both forearm vessels	Always survive with appropriate pre-hospital treatment

Source: Davies and Campbell (1962), as modified and reported by Caldicott *et al.* (2001).

triangular bite); signs of crushing are uncommon (White 1975). Sharks with finer serrated tooth edges produce sharp wound edges, while species with coarse or minimal tooth serration result in deeper puncture wounds. Many large shark species have a combination of sharp, pointy bottom teeth, used to grasp flesh, alongside more triangular-shaped,

Table 7.2. Level of shark attack injury, based on the sum of the scores from the SIT scale

Level of injury	Total SIT score	Description
1	0–2	Simple lacerations involving the skin and soft tissue; blood pressure is typically unaffected; loss of function is not seen
2	3–4	Skin and soft tissue injuries that tend to involve a muscle, tendon or bone; patients are quickly stabilised without much blood loss; function of extremity is not compromised
3	5–6	Complex lacerations that typically involve muscle, tendon or bone; patients may have transient hypotension and loss of function of a tendon; they likely require future surgical procedures for adequate repair of the wounds
4	7–8	Aggressive attacks that result in deep tissue damage and loss of function of an extremity or organ; a major vessel is likely to be injured; patients are hypotensive and require immediate surgical intervention to prevent fatality
5	9–10 (fatal)	Most likely a fatal injury resulting from the severity of the bite, hypotension, loss of function of an extremity or organ or rapid blood loss

Source: Modified slightly from Lentz *et al.* (2010).

serrated top teeth used to saw. This tooth structure, in conjunction with rapid and forceful head-shaking, is designed to tear not to chew, which means that injuries from those species characteristically involve areas of stripped flesh (İşcan and McCabe 1995; Caldicott *et al.* 2001). The sheer bite force of many large species is sufficient to bite through bone. Most surface swimmers are bitten on the lower limbs only; however, upper limbs are often subsequently injured when victims try to fight off their attacker (Burnett 1998; Caldicott *et al.* 2001). Although potentially quite severe, injuries to the extremities are often non-fatal. They do, however, require early intervention by reconstructive surgeons to salvage what is left and to cover the remaining vital structures, such as bones, arteries and nerves (Ricci *et al.* 2016).

Whether alone or in combination with wounds to the extremities, injuries to the torso pose a greater challenge to surgeons (Ricci *et al.* 2016). These wounds result in more fatalities, yet some patients do survive. Torso injuries include tissue loss from the abdominal or thoracic walls and potential organ exposure. Wound closure is complicated, as torso injuries are more likely to include areas that are normally used for donor tissue (e.g. the back) for wound repair.

Surgery

Often, the patient is already stabilised in terms of haemorrhage and fluid volume by the time they arrive at the hospital. Therefore, when they reach the emergency department, they are cardiovascularly stable and can go straight into surgery. However, if this state has not yet been achieved, their instability must be rectified before any further investigation or reconstructive action can begin. A multidisciplinary approach to injury management is necessary for optimal results. Typical treatment occurs over numerous stages. An initial assessment of deep structures, such as tendons, vessels and peripheral nerve injuries, is made, and photographs taken. Plain film radiographs are taken to identify any fractures or residual tooth fragments, which could become a source of sepsis if not identified and removed (Caldicott *et al.* 2001). The injuries are swabbed and sent for culture so that more informed antibiotic therapy can be prescribed, then thoroughly washed to remove microbiologically contaminated particulate matter, such as sand and

I treated one patient who came in with a horrendous injury from a shark. He arrived in a helicopter quite soon after the attack. He had already been intubated and ventilated and was in an induced coma by the time I saw him. What I feel was quite important, in terms of his survival, was that they had recently made the decision to have blood available on the helicopter. He had tourniquets on his legs, which I think was also quite important. As a rule, we don't like tourniquets, because they can do quite a lot of damage. But in some cases you have no other choice to stop the bleeding, and you just have to get the patient to help as quickly as possible. This injury was absolutely lethal. Luckily, he didn't have a significant arterial injury. If he had, he absolutely would have died. I guess if you have a significant shark attack, it's a big beast, and it makes a big hole in your leg or wherever it hits you. And if it gets your torso, then that's the end of it, I can't see how you'd survive that. Blood loss is the main problem with these things. The power involved in this injury was unbelievable.

Most of what needed to be done involved cleaning the wound, tidying it up, stitching it all together and repairing the structures underneath. Initially, the team was a bit shocked and it took a moment to get them to focus and get their attention on what we had to do. Ours was the first surgical team involved, our task was to control the bleeding. Vascular surgeons sometimes are asked to be involved when there is a lot of bleeding. We were not involved terribly much after that but I believe they were washing sand out of the wounds for days afterwards. I imagine that is a distinct difference in this type of trauma.

With shark attacks, at least going off the cases that we saw here, the patients tend not to actually be missing a lot of tissue. With our patient, there was a massive semi-circular flap of skin that was raised from one leg, but none was missing. This shark's jaw must have been at least 18 inches across; it would have been a massive animal. If he had been attacked with a butcher's knife, it would not have done that much damage. It was absolutely impressive. But it wasn't like the shark took a big chunk of the guy, and then left. My understanding is that what happens is that the shark takes a bite, to taste really, only to realise that it's a case of mistaken identity, and then pushes off. The shark had a mouthful of this guy, so why didn't it swallow it? It was obvious to me that it took the bite and changed its mind.

I guess shark injuries are in some ways similar to any major injury, but the truth about trauma normally is that often the victim carries some of the blame for the injury, possibly driving too fast, not wearing a helmet, or doing something that is not sensible. Tempting fate in some way. Alcohol is often involved. Even gunshots, they are very rarely a random thing. To me, shark attack is different, and I guess that's why I feel differently about it. I mean, here's a guy who is just getting on doing what he loves doing, just surfing or going for a swim in the sea. And while I don't have a great passion for that sea, I still understand passion. And I accept that for a lot of people being in the sea is a big deal for them. These victims are essentially just good people getting on doing something they love, and suddenly they are in a world of trouble. And that is why it is different, they are innocent.

People often say that the guy that I treated for shark bites was lucky to save his leg, or to be alive. I am not sure that getting bitten by a shark could ever be described as lucky, really. But from our perspective, it was a happy story. The guy was pretty messed up, but he came in to us and got fixed up, and he walks again. He has a nerve injury, so it's not perfect, but he's up and he can get about. And while it's not perfect, considering that it could have easily been a fatal injury, it's pretty good really.

In cases where the outcome is not so good it can really have an effect on the team. You invest an enormous amount of energy and emotion to get these patients up to the operating theatre and get going. If it all falls apart and the patient does not survive, it can be quite upsetting. But as professionals in the field, I think that we just internalise it, and get on with it. We don't have stress counselling for this; maybe we should. And even though we had such a good outcome for this patient, I was still affected by it. I had to engage with this particular case more because it was filmed for the television show *Gold Coast Medical*. As a result, I was forced to actually think about it a lot more than I might have done ordinarily. Normally I would have just gone 'Holy mackerel', and then moved onto the next job. But because I had subsequent interviews with the TV crew, I was forced to think about it a lot more and engage with the situation to a much greater extent. A lot of people wanted to know what I thought and my opinion, and so on.

In terms of the effect on me, it's unlikely that I will ever swim in the sea again, just from seeing that one shark attack victim. And that is different for me. We used to love swimming in the sea at Durban in South Africa when I was a child. And my family and I would swim at the beach here on the Gold Coast a couple of times a year. But since that case, we haven't been in the water. I'm not a hysterical person, but I am rational, and I just think why would you mess with that? Because if a shark attack does happen, it's a cataclysmic injury. But at the same time, life is about risk, it's a fantastic thing. We tell kids not to climb trees because they might get hurt. But which is worse, a child falling out of a tree, or that our children don't climb trees? I don't profess to know all the answers to these questions but a world without risk would be a very dull place indeed. But to my mind, if children don't fall out of trees, then they never understand consequences. Sometime bad stuff is going to happen, and we have to accept that. If sharks are a big issue to you, then get out of the water.

Overall, we are getting better at medically managing trauma, and that would include shark injuries. But most shark attacks occur in relatively remote places, which is highly significant. Shark attack victims might be alive at the beach, but they don't survive the next hour. So it doesn't really matter if you get the technology right and everything else is going perfectly, if the surgeon is not standing close by with a team to help and an operating theatre handy, then it's going to be a problem. There are components that help to buy time, the whole resuscitation process and stabilising the patient. But ultimately, if there is a hole in the tank and you want to fill it, you need to close the hole.

If you come across a person who's been attacked by a shark:

- Do what you can to control bleeding, we prefer people to put manual pressure on bleeding rather than a tourniquet. Use towels or clothing. If it is a limb, a tourniquet may be necessary, use a rope or something similar and wind it firmly around the leg or arm above the injury.
- Try to cover the wounds up as best as you can, and prevent contamination by the sand.
- And then initiate getting help. The priority is to move the person out as fast as possible to a place where they can get advanced treatment. That's the real priority – moving them. You have to get help. There is not a lot of point spending ages on the beach, because that is not going to get the job done.

Dr William Butcher
Vascular Surgeon, Gold Coast University Hospital

debris. Dead, contaminated and all associated tissue is completely removed. Severed nerves can then be repaired. Fractures should be assumed to have been contaminated by the external environment, and treated as such. Finally, a plastic surgeon closes any remaining sites over a drain or packs them with dressings to allow for delayed closure. Abrasions and small lacerations or punctures should be thoroughly irrigated, and topical antibiotics applied (Burgess *et al.* 1997; Caldicott *et al.* 2001).

Microbiological pathology

Although haemorrhage is the primary concern and cause of death following major shark attack trauma, the risk of secondary infection in shark bite injuries of any degree should not be underestimated or ignored. With the increased percentage of shark bite victims surviving the initial stages of resuscitation and trauma management, the management of secondary infection has become a major focus. Aquatic environments are in no way sterile, supporting a wide array of atypical bacteria capable of infecting humans through open wounds. Similarly, shark teeth support a variety of bacteria. Additionally, injuries are often subject to other environmental contaminants, such as sand, and early treatment of such wounds is essential. A few really interesting studies have cultured the bacteria present on the teeth or in the oral cavities of sharks, including white sharks (*Carcharodon carcharias*), bull sharks (*Carcharhinus leucas*), tiger sharks (*Galeocerdo cuvier*) and blacktip sharks (*Carcharhinus limbatus*) (Buck *et al.* 1984; Interaminense *et al.* 2010; Unger *et al.* 2014). Cultures isolated up to 81 potential bacterial pathogens in individual species. The majority were Gram-negative enterobacteria but several others, most notably *Vibrio*, *Pseudomonas*, *Staphylococcus*, *Enterococcus* and *Streptococcus* species, were also isolated. Although further testing revealed that not all were zoonotic (pathologically transferrable from sharks to humans), some were thought to be potentially problematic to bite victims. Both *Vibrio* and *Pasteurella* species are pathogenic, and can cause life-threatening infections in humans (Davies and Campbell 1962; Pavia *et al.* 1989; Royle *et al.* 1997; Caldicott *et al.* 2001; Woolgar *et al.* 2001; Rtshiladze *et al.* 2011) and *Pseudomonas putrefaciens* has been associated with septicaemia from wounds acquired in sea water (Roland 1970; Barker *et al.* 1974; Hugh and Gilardi 1980).

The extent of shark bite lesions has been shown to be the main contributing factor in the degree of bacterial colonisation. Although bites from smaller sharks, such as blacktips, may not lead to major traumatic injuries and are rarely fatal, this category of sharks is responsible for the highest prevalence of bites on humans and even minor bites present an opportunity for bacterial transfer from sharks and sea water to the human recipient. The resulting potential for infection and related complications means that timely wound management and appropriate antibiotic therapy are critical.

Antibiotic sensitivity profiles suggest that while many bacteria are resistant to one or multiple drugs, effective initial preventative antibacterial therapy could begin with a single broad-spectrum antibiotic drug, but special attention should also be given to the more acute management of Gram-negative bacteria known to be associated with sharks and sea water. Culture and sensitivity studies of the bacteria isolates from shark teeth have been supplemented by the medical records of shark attack patients treated in hospital. These studies suggest that empirical antibiotic treatment should focus on *Vibrio* spp. and, although not identified in any of the tooth culture studies, *Aeromonas* spp. More generally, staphylococcal and streptococcal infections should also be assumed and treated, and prophylactic chemotherapy for tetanus should be given on the patient's arrival at hospital (Caldicott *et al.* 2001).

Despite early prophylactic treatment for bacterial infection, culture and sensitivity profiles from each patient's wounds should inform and direct more tailored ongoing treatment. Patients must be closely and continuously monitored for signs of abrupt and severe recurrent or new infection, such as rapidly progressing cellulitis, myositis or necrotising fasciitis, which can lead to critical infection within hours of exposure (Royle *et al.* 1997; Caldicott *et al.* 2001; Unger *et al.* 2014; Ricci *et al.* 2016). Wide-scale debridement of injuries during initial surgeries can alleviate some of the risk for subsequent infection. However, severe infection may require subsequent surgeries for further site debridement.

Interestingly, and of significant environmental management concern, is that notable human pathogens isolated from blacktip sharks in Florida, including *Staphylococcus aureus*, *Pseudomonas* sp., *Shewanella putrefacians* and Enterobacteriaceae (e.g. *Enterobacter* sp., *Klebsiella* sp., *Escherichia coli*), may be a result of the animals' exposure to human sewage effluent (Unger *et al.* 2014).

Post-operative care and concerns

Shark attack victims may be subject to extensive and prolonged post-operative treatment and management. The development of renal failure is common following significant hypovolaemia, myoglobin release and the compounding neurotoxicity of some antibiotics (Guidera *et al.* 1991; Caldicott *et al.* 2001; Ricci *et al.* 2016). It is also not uncommon for patients of major traumas to require multiple operative procedures and transfusions, which means that the potential for coagulopathy (an impairment to the blood's clotting ability) should be assumed and anticipated.

I have been practising in the field of trauma surgery for 20 years, and have treated a wide variety of injury patterns. However, the first major shark attack trauma I saw was last year when I started working for the Gold Coast University Hospital. And within that one year, we had three major shark attack victims. Since coming to Australia 10 years ago, I've seen smaller wounds from smaller sharks, but they were just skin lacerations and flesh wounds to arms or legs. Normally those patients cope and recover well and might not even need to stay in hospital that long. But the three we had last year were major trauma injuries with large bleeding flesh wounds. Interestingly, all three victims (all surfers) were bitten on the back of the thigh. So the sharks obviously came from behind when they stood on the board and picked the thigh – the largest fleshy part! I don't know if that's on purpose or not, but fortunately, that location is also the furthest away from the major vessels in the groin. The big sharks would go for the largest area, I suppose. One patient also had injuries to his arm from fighting with the shark. All three cases were attributed to white sharks, but the hospital was never asked to assist with identifying the species in any way.

Overall, treating shark attack trauma is quite a horrific experience for everyone involved: for the patient, the bystanders, the paramedics, the pre-hospital rescue physicians that attend at the scene, the transport team and the emergency doctors on arrival at the hospital. It really affects everyone involved, particularly those involved early on. I don't want to say that people become scarred, that's not really the right term, but those horrific big wounds impact a lot of people. There has been a lot of talk within the hospital, particularly from those who were there at the patient's arrival or down the track in the operating theatre, which has had an interesting dynamic. These three events definitely impacted the hospital, as well. Because of the rarity and emotional aspects of shark attacks, they get a lot of traction. The media were involved, and stories got to the radio and TV, ramping up the momentum, and the hospital was given titles, like the Shark Centre.

I wouldn't consider shark attack traumas that different to any blast injury or bigger explosion injury. There are some different facets to shark attacks, but ultimately, treatment is about assessing the patients and treating them according to the injuries and life-threatening priorities. As a trauma surgeon, you see a lot of big stuff throughout your career – horrific injuries, bleeding, and gut-dispelling injuries. You don't ever truly become used to it, or blunt to it, but it's nevertheless part of our professional job. But the biggest difference in major shark attacks is the bleeding. This is where you save lives, or where the patients die. It all comes down to whether you can stop or control the bleeding early on, together with sufficient resuscitation and giving blood products. We see motorbike victims every day, but the tension, traction and speciality of shark attack trauma is different. I suppose something different is the type of wound. If it's a big attack, then you have a flesh wound that is more of a grabbing, ripping, shearing type of injury, with more or less tissue loss, and you end up with bleeding from the tissues and muscles. That's even worse if the big blood vessels, like the big arteries or veins, are involved, and in these cases it is so important that you control those big vessels first. If a shark happens to open up your main leg artery, then you are very unlikely to survive.

The biggest surgical challenge with the shark attack wounds that we saw was the shredded, bleeding soft tissue flesh and muscle damage. Muscle tissue is highly perfused, with good circulation, so if a large chunk of it is removed with an open surface or cavity, then massive bleeding can occur. The plan of surgical attack is to control the bleeding by putting pressure on the main artery (called 'proximal control') to stop the blood loss and to get a better overview of the real bleeding sites. After identification of these small bleeding vessels, they can be sacrificed and tied off. The temporarily compressed main vessels are released and the bleeding re-evaluated. If it has stopped, then that's a big relief for the team and (obviously) the patient. On the other hand, if bleeding continues, then further manoeuvres need to be performed until bleeding is controlled.

There are different stages to trauma, each with different risks. For shark attack, initially, it is the exsanguination, or bleeding out. If that can be controlled and the patient passes that stage, the next stage includes a couple of risks, first of which is the concern for serious infection. Infection is common in animal trauma, and it's a big deal, particularly with certain animal bites such as dogs, cats, monkeys and others, including sharks. The bacteria in the mouth of animals is quite aggressive and nasty. It gets into the tissue, through tooth puncture wounds. These punctures may not result in large open areas, so can be fairly difficult to wash out. Shark attacks also come with other sources of bacteria, such as sea water, which is not very clean, and sand. One of the victims had crawled out of the water through the sand, so altogether you can end up with highly contaminated wounds, which might cause serious infection. The second subsequent stage of concern for major traumas where there is significant damage to soft tissue is the systemic inflammatory response syndrome. This can occur when micro-material and molecules from the damaged tissue get released into the bloodstream, causing an inflammatory response of the body, which shuts down the immune response, circulation, leading to organ dysfunction including the ability for gas exchange in the lungs or kidney failure, and ultimately multi-organ failure fighting for life.

It has been interesting to see the emotional transition that shark attack patients go through. It often includes stages of anxiety, denial, nightmares, and they get quite emotional talking about it. Then it's the acceptance of not just that it happened, but also of the injuries. And the injuries do not just include the horrific major injuries. There is also often permanent damage – scars at the very least, and missing flesh, which doesn't grow back. It can also be a real mental battle. And that is not only for the victim, but also the family, friends and other involved people. Depending on the injuries, you might also have to discuss potential limb amputation with the patient. And that obviously lifts the emotional battle to an even higher level. The surgeons ultimately make the decision to amputate or not, but we definitely involve the patient and the relatives in this process. When trauma like this does occur, it hits rock-bottom.

In terms of the most advantageous response for bystanders who may be the first responders to a shark attack victim, it would have to be basic and advanced life support. The bleeding needs to be controlled and stopped. With major bleeding, the best approach, after applying local compression, is to tourniquet the limb above the injury. Proper tourniquets are not absolutely necessary, you can just put a strong string, or whatever may be on hand that can be used to improvise, around the limb to tie it off to block the blood flow. However, that is very difficult if the injury occurred in the armpit or the groin, in which case the best approach is simply through the application of pressure. But injuries to those areas are quite often fatal by nature.

Associate Professor Martin Wullschleger
Medical Director of Trauma, Gold Coast University Hospital

Rehabilitation may be extensive and multidisciplinary. It is likely to include, along with more patient-specific treatment, ongoing plastic surgery and physiotherapy. Additionally, the psychological effects of the attack and resulting injuries to the shark attack victim and their family and friends should not be understated (Davies and Campbell 1962; Charlesworth 1976; Caldicott *et al.* 2001).

Psychological aftermath of shark-inflicted trauma

Psychological trauma is a typical response to significant shark bites; however, it can result from any event in which a person experiences, witnesses (or is first on the scene) or learns of someone close to them dying or being threatened with death, serious injury or violence. Traumatic events are often highly isolated and overwhelming experiences. Usually, they're like nothing a person has ever experienced previously, and they can be difficult to come to terms with. This is especially true when debilitating physical injuries are involved, as pain and psychological distress are inextricably linked and may be mutually reinforcing (Phoenix Australia 2017). Trauma is often accompanied by strong feelings of fear, sadness, guilt, anger, grief, despair and/or loneliness. We are psychologically prepared for car accidents and motorcycle crashes but not for shark attacks, which are much less common. The rarity of shark attacks may also exacerbate the reactions that people experience, due to a lack of community understanding (R. Cash, *pers. comm.*).

While most people recover from trauma quickly, especially with the help of family and friends, others may feel long-lasting, complex and overwhelming effects, leading to compounding mental health issues, such as post-traumatic stress disorder (PTSD), depression, anxiety, or alcohol and drug abuse. Personal relationships with family, friends and work colleagues may be impacted, making it even more difficult to come to terms with the trauma. The way people are affected by, and recover from, trauma depends on a variety of factors. These include the degree of terror or horror in the event, how unexpected or out of the ordinary the event was, and how much perceived control a person had (R. Cash, *pers. comm.*). During a traumatic event, an individual's level of control over the situation affects their resulting level of anxiety. Situations where a person is able to escape, or even fight back, appear to lessen the likelihood of developing a lasting fear of the event (Mineka *et al.* 1984). If an event is inescapable, however, and a complete lack of control is experienced, the experience is far more likely to result in a greater degree of conditioned fear and trauma. If another reason to fight back in the case of shark attack were necessary, there it is.

Post-traumatic emotional distress can be so intense that the situation can be perceived as unmanageable, impairing functionality, relationships, normal day-to-day life and the ability to relate to others. The latter may result in withdrawal from family, friends and social activities. Isolation is commonly linked with PTSD, and the feeling of isolation was repeatedly mentioned by those affected by shark attack. Feelings of anger that persist more than a couple of weeks after a trauma can seriously impair personal relationships. Anger has been found to exacerbate PTSD change over time and to decrease the success of treatment (Phoenix Australia 2017).

Conclusion

Our ability to respond to and treat shark bites has improved appreciably. This is explicitly demonstrated by the significant reduction in shark bite fatalities over the past century. Although most shark bites are relatively minor, the environmental conditions where bites are sustained and the nature of the bacterial fauna within sharks' mouths make it imperative that all shark-inflicted injuries are seen by a medical professional for maximal recovery. Although there are condition-specific intricacies, the treatment of a major shark attack is no different from that of other traumas. The most urgent response is the control of bleeding, and this is often done at the beach with direct pressure to open wounds or by tourniquet. Another major component of shark bite treatment is the management of secondary infection. Shark attacks may be not only physically traumatic but also mentally traumatic, and the ongoing management of post-traumatic mental health is extremely important for shark attack survivors.

8

Human–wildlife conflict and regional management

Human–wildlife conflict is a significant and widespread global issue that relates to human safety, tourism, emotion and the feeling of freedom of choice. It also affects animal welfare and conservation, as conflict generally leads to a devaluation of wildlife. The diversity of wild animals that conflict with humans in some way worldwide is huge, and includes an assortment of predators, herbivores and insects. Conflict may arise due to overlapping resource needs or more directly, where there is potential for people to be harmed by wildlife as a result of predation or protective behaviour. Worldwide, large carnivores were responsible for the deaths of an average of at least 150 people per year during the 20th century (Löe and Röskaft 2004). Despite the diversity and global occurrence of human–wildlife conflict, few if any species are more feared than sharks. Shark–human interaction is the most geographically dispersed human–wildlife conflict in the world (Crossley *et al.* 2014).

Human–wildlife conflict is increasing globally, as expanding human populations move further into previously uninhabited areas. Also, as conservation actions take effect, previously depleted wildlife populations are beginning to reinhabit their natural ranges in greater numbers. A range of mitigation measures has been developed for a variety of species that present risks to human life or livelihood, including physical separation of conflicting species and resources (enclosures, repellents, deterrents), guarding assets (physical devices or human guarding), habitat use and modification (habitat zoning or modification to limit human–wildlife overlap), behavioural modification of conflict-causing species (physical or taste aversions), behavioural modification of humans (education or human relocation), use of buffer resources (artificial provision of alternative food sources), lethal control of conflict-causing species (culling, retaliatory killing), non-lethal control of conflict-causing species (relocation) and reducing the cost of conflict (reducing economic losses, providing economic incentives or alternative income possibilities, or increasing the economic value of problem species) (Dickman and Dickman 2010). While not all of these would be relevant to potentially dangerous shark species due to their global distribution and effectively limitless range, a surprising number are.

Do we need shark attack mitigation?

Sharks and shark attacks are undeniably controversial topics and have become increasingly popular in the media in recent years. It is, therefore, not surprising that surveys of public opinion have been conducted to gauge public views on these subjects. A UK-based study found that, of respondents who had encountered sharks in the wild (in various parts of the world), less than 5% described their encounters as negative and none of the respondents carried negative feelings towards sharks based on their encounters (Friedrich *et al.* 2014). Overall, those who had encountered sharks in the wild tended to have more positive, pro-conservation attitudes than those who had not. However, although few people have had negative encounters with sharks in the wild, the widespread fear of shark attacks means that the footprint of such an event has the potential to be much greater. In certain regions, where attacks occur (or are perceived to occur) with higher frequency, entire economies can be affected. Thus, mitigating the potential for shark attack has become a focus of regional and national governments around the world, instigating the conceptualisation, trialling and deployment of mitigation measures over the last half-century. Broadly, shark attack mitigation measures are actions taken to reduce the prevalence, risk and perception of the likelihood of shark attacks.

A successful response to the threat of shark attack (sometimes including the development and implementation of shark attack mitigation measures) involves several challenges.

1   The decision of whether to act at all. This is challenging due to the disproportionate fear of, and highly emotional response to, shark attacks among the general public. Such irrational fear can place a great deal of pressure on governments to take action in situations that, based on statistical relevance, would otherwise not need to be addressed. Conversely, the decision must factor in the possibility of grave consequence (fatality, major trauma, psychological damage) in the event that an attack does occur, and the high media involvement and publicity surrounding incidents. The question of whether to take mitigating action in relation to shark attack is very much a regionally specific question, and the overwhelming answer for most locations is probably no. For example, a 75 km stretch of Atlantic Ocean beaches of Volusia County in Florida contributes heavily to the annual occurrence of shark attack (up to 30% of global annual bites). Volusia County is often given the title, 'shark bite capital of the world'. Yet, with the exception of signs which state that hazardous animals have been seen in the area, no mitigation measures are employed there, and the general feeling is that there do not need to be any. Various regions in Australia, South Africa and Réunion Island, which typically have smaller numbers of bites but a much greater percentage of serious and fatal incidents, are far more justified in employing mitigation measures.

2   The dispersion of potential attack locations, under different environmental conditions, and the wide range of activities undertaken at the time of an attack. The ocean is a big place, and the potential location for a shark attack is just about any natural (or naturally connected) waterway where humans and sharks overlap. Thus, there is no way to provide complete coverage against shark attacks, unless all sharks that pose any sort of risk to humans are removed from the natural environment. Attacks occur on swimmers and waders pretty much as close to shore as you can get, or in the middle of the ocean. They occur on surfers in heavy sea conditions, in remote areas away from patrolled beaches, on isolated commercial abalone divers, on spearfishers at depth – and on groups of paddlers in protected bays. Again, it is

impossible to protect all water users due to the vastness of activity location and environmental conditions. Linked to this challenge is the technological difficulty of designing and deploying regional mitigation devices in such dynamic environments, which are continuously subject to the unpredictable and inconsistent forces of nature. And, going back to the fact that the ocean is a big place, any sort of deterrent device would need to function in an effectively limitless open-water environment that is continuously affected by currents and rapid dispersion and that is, in general, resoundingly different from our far more broadly and intricately understood terrestrial world.

3   How to effectively protect humans while not having adverse environmental impacts. Several shark attack mitigation measures have the potential (and have been undeniably shown) to have major detrimental impacts on natural shark populations, including some that are of conservation concern. The potential adverse effects of mitigation methods on the environment and other aquatic life needs to be taken into account. There may be effects on harmless shark or ray species or other aquatic animals, and the destruction of underwater flora or structures. Hazardous chemicals may be deployed and dispersed into the water; benthic flora, fauna or structures may be destroyed by nets; and noise pollution has the potential to alter the natural behaviour of other receptive species. Public awareness and perception of our environmental responsibility has increased and now, more than ever, the public demands that governments be held ethically responsible and accountable for their actions. Questions regarding the environmental effects of proposed methods and demands for publicly available information on the efficacy of those measures are at the forefront of conversations. These may relate to the amount of bycatch a particular method results in, or emissions from manned or unmanned aerial patrols.

4   The significant interspecific variation in shark biology, physiology and behaviour. Shark sensory systems can be extremely variable, as they have evolved to support the specific habitat and behavioural patterns of the species, and their predation and

predator avoidance strategies. The discrepancy between species can be so great that what might repel one shark could attract another. This presents a highly relevant challenge for mitigation in locations that have a wide diversity of shark species, such as the east coast of Australia.

5   And of course, the cost of mitigation is important. The cost of designing, trialling, deploying and maintaining new or existing technology, built at a scale that would present a real chance of deterring sharks, in often very tumultuous marine environmental conditions, can be exceptionally high. The financial decision to take mitigating action against sharks can be especially complex from an ethical perspective. When considering the statistically low probability of shark attack, the decision to allocate large sums of money to the situation would be difficult (especially for measures that may be unproven, or have been shown to be equivocal in effectiveness) when there are many more hazardous or lethal human health and safety situations and conditions (e.g. cancer, drownings, road traffic accidents) that may also require funding from the same pool of money.

The decision-making process about the allocation of funding to shark attack mitigation measures needs to really scrutinise the technology investigated. Savvy entrepreneurs are keenly aware of the impact of public perception and the human tendency to err on the side of safety (the precautionary principle), which suggests that people call for safety measures even when cause-and-effect relationships have not been fully established (Wilson 2010). Thus, our lack of understanding of shark behaviour, motivation and movement patterns offers openings for entrepreneurs to provide 'solutions', even if those solutions involve little proof of concept, evidence of effectiveness or verification that they do not involve unintended repercussions (Wilson 2010; Neff 2012).

To kill or not to kill

One of the main questions in regard to shark attack mitigation is whether a reduction in shark populations as a whole, or the removal of particular individuals, would actually reduce human fatality or serious injury. This is especially questionable given that shark numbers have been shown to be unrelated to shark attack rates, and that population reduction measures have not been proven to be effective in terms of shark–human interaction mitigation. Furthermore, the removal of one predatory species does not necessarily equate to the removal of a threat to humans. As stated by Larkin (1979), 'don't expect big changes for a predator that loses a species of prey … or for that matter for … the prey that loses a predator. Don't expect long-term benefits to prey from predator control.' Gibbs and Warren (2014) polled 511 Western Australian ocean users, including surfers, board riders, divers, snorkellers, swimmers, paddlers, fishers and lifesavers, and found that 69% had encountered a shark. The list of species encountered was extensive, but the three species considered to be most dangerous to humans and most commonly targeted in lethal mitigation programs (white, tiger and bull sharks) were included. Indeed, of the respondents who were able to confidently identify the species they had seen, a substantial 54% had seen tiger sharks (*Galeocerdo cuvier*), 23% white sharks (*Carcharodon carcharias*) and 20% bull sharks (*Carcharhinus leucas*). This suggests that water users encounter sharks, including those considered potentially dangerous, on a regular basis, yet the overwhelming majority of encounters do not have a negative result. This frequency of harmless encounters is generally overlooked when policy is considered, and the representation of negative shark–human interaction is heavily distorted by the few encounters that do result in bites.

However, the idea that 'the only good shark is a dead shark' is still held by many people, regardless of the science, statistics or conservation status of the animals. Lethal methods remove animals from the environment but in most cases, even for individuals of potentially dangerous species, it is unknown whether a particular animal ever had, or would have, bitten a human. Thus, the removal of large numbers of animals on the basis that a very small minority may be dangerous to humans, is often seen as extreme. Tiger sharks represented most (91%) animals caught in the Western Australian shark drumline trial in 2014. While tiger sharks over 2 m in total length were a target species of the program, the reason for that was heavily questioned. Statistics from the Global Shark Attack File show that there have been only four attacks on humans by tiger sharks in Western Australian waters: two in the 1990s were fatal, one in January 2012 resulted in lacerations to an arm, and one in December 2012 did not result in an injury.

Direct shark population control measures, due to the perceived risk of these animals to humans, fisheries, tourism and occupational safety, have contributed to the threatened conservation status of at least 12 shark species (Dulvy *et al.* 2014). Many mitigation strategies either focus on or include white sharks, bull sharks and tiger sharks. White sharks are listed as Vulnerable by the IUCN, with an unknown population trend, and are protected in various regions worldwide. They are listed in CITES Appendix II and are considered, by some accounts, to be facing a high risk of extinction in the wild. Bull and tiger sharks are both classified as Near Threatened. Due to the protection and conservation status of white sharks in particular, lethal programs, or those with the potential to be lethal, may need special permission from various legislative bodies before they are allowed to proceed.

Lethal shark attack mitigation measures are not without irony, and legislation and policy relating to shark attack have revealed a 'predator policy paradox' (Neff 2012). Where governments feel forced to act, there are few political incentives to uphold policies that protect endangered species, if those species may present harm to the voting public. The acceptance of killing endangered species is thus validated on the basis of easing public fear. Australia, which was the first country in the world to pass legislation protecting a shark species and which has since afforded protection to several other species, including the white shark, nevertheless employs indiscriminate lethal mitigation measures in various regions. National protection for some shark species means that any measure that threatens to kill, injure or move those animals must be done under permit, and any action that could have a significant impact on one of those species must be considered by the Department of the Environment before commencement. White sharks are the only shark species that has been given full protection in South African waters, yet lethal mitigation methods for the species are employed. Similarly, Réunion Island implemented a ban on commercial shark capture in 1999 and banned fishing entirely along a 40 km stretch of

I wasn't actually there when Doreen was attacked, but I spent time with her husband on the day, and I went with him the next day to identify the body. Her death was just so shocking. I imagine this is true for all unexpected and tragic deaths, but you don't easily come to terms with it. Doreen was a very special lady and she is really missed.

People often say don't kill the sharks, but if there is any other animal that attacks or kills someone, they kill that animal. If a dog attacks someone, they put the dog down. To me, there is no debate about not killing that shark. This isn't to say that I want all sharks gone; I don't, but I did want that one killed. Actually having someone close to you killed by a shark makes you think differently about things, I guess. And some might say that that is an emotional response, but we just don't know enough about their behaviour. Are they more likely to do the same thing again? Will that same animal kill someone else? You just don't know, but is it worth the risk? The timing of the political response needs to be better, though. If you want to catch a shark that killed someone, you can't have 24 hours of debate first to make the decision, like in this case. The response simply wasn't quick enough. I think they sent the boats out to look for the shark the next day, and then they said the weather was too bad, so they brought them back in. People deal with things differently, but when you are very angry, you want to see material things happen. And once you've lost someone, it becomes personalised.

There are a lot of people here who spend a great deal of time in the ocean, like surfers and boaties, who are saying that there are a lot more sharks in the ocean, and that they are a lot closer in. While this isn't based on scientific analysis, it's all anecdotal, these are the people who are actually in or on the water all the time. And you can get a real sense of things changing before you can show it in hard data. But I don't really feel that that is being picked up on. Science says shark attacks are rare, but if you want to go and have a swim at the beach, you don't care, you just want to feel that there is something there to protect you.

Sharks are regularly in the area and of course they're in the ocean, it's their ocean. We've always known they're here. I suppose now it's just a matter of them coming closer to shore. It doesn't feel like they're just out in the deep anymore. A lot of swimmers are now joking that they are getting gravel rash on their fingers because they are staying so close to shore as they swim up and down the coast. There seem to be more whales, and perhaps there are more sharks because of that, but the whales aren't 50 m off the beach, so that can't be the whole reason that sharks are now so close to shore. Perhaps it's the same few sharks terrorising the Perth beaches, but the sightings have certainly increased. There is a network of buoys that record sharks in the area and you can download data from that, but I never look at it. I figure if I did, then I would never get in the water again. One of the people that I swim with looks at it all the time, but she still swims in the ocean every Sunday morning; it hasn't changed her behaviour.

We used to spend two weeks every year on the coast in Victoria, and we'd go into the water. Occasionally the shark alarms would go off, and everyone would get out. A speedboat would go out to chase the shark away and then everyone would pile back in. And no one thought any more of it. But things are different now. Western Australia has had a lot of sightings this year (2016) and we've had two attacks so far this year, both fatal. Our beaches have been closed for a month or more at a time, even over the winter, because of shark sightings, and that isn't typical for us here. So I feel that something must be out of balance somewhere in nature – whether we've overfished and sharks are following the smaller fish in so they are now closer to people, or if we've protected the sharks too much and upset the normal balance in nature. But something has changed and, from that perspective, I do feel like something needs to be done. Yet I feel that the political response has been quite soft, and they are just arguing about what to do. But what is actually best for the community? You can't stop shark attacks from ever happening again, but you would hope that we can manage the issue better going forward. I think you just have to do something; try what you think might be most effective and then evaluate it properly, but don't just argue over it. I admit it's a difficult issue, and I wouldn't consider myself an expert on the topic, but I'm not convinced that we're doing enough and I'm not confident that we are handling it well at the moment. I know they use drumlines in Queensland and there have not been many attacks there, so I think that this could be a reasonable step. I don't think that they should be deployed every-where, but there are several beaches in Perth that are very popular in the summer, and for those beaches I think it would be a really good idea for people to have some sort of reassurance. And if they catch harmless sharks, then put them back, but if they are potentially harmful, then I would say tag them and take them somewhere and see what they do. Do they come straight back? If they do, then you've got something more to think about.

In terms of resource allocation, it's a bit like gun control. We don't have that many mass murders, but it's the shock and horror of what can happen that makes you want more done. If you looked at shark attack mitigation as an intervention to keep people safe, depending on what that intervention is, it could be as cost-effective as many other things that society pays for. And this would be better than trying to manage and pay for something after the fact. If people stop going to the beach, then there are a lot of other businesses and activities that will suffer as a consequence. And if things keep going the way they are, then I believe that this is exactly what will happen. A lot of people are already thinking twice about going to the beach after what's been happening on our coast in the last few years. I have friends in the surf lifesaving clubs, and they would say anecdotally that the numbers of people going to the beach are already reduced. And parents are pulling their little kids out of Nippers pro-grams [introduction to surf lifesaving skills, for children] left, right and centre.

I don't think that Doreen's death could have been prevented. They were 2 km out on a reef that they had dived hundreds of times before. But if some of the broader strategies in place were more effective, then maybe the coincidence of that shark and her being in the same place at the same time might not have happened. Doreen and her husband had talked about devices that people can wear to repel sharks, but this shark was 5.5 m long, and I don't know how well anything would repel a large animal like that. These devises are expensive, their efficacy is still in question, and considering she had been diving for ages and had never seen anything to be concerned about, she had decided not to get one.

I'm an open-water swimmer, and had always gone to the beach for swims. But now, if I want to go for a swim, I'm more likely to go to a pool. I know there are lifesavers around and there are shark-spotting programs, but I just don't feel as comfortable. I'd certainly never swim in the ocean by myself. I'll still do organised events in the open water, because I figure there are enough boats and surf lifesavers and planes, so my risks would be minimal. But I don't particularly enjoy going for a swim in the ocean anymore. I am going to Rottnest Island next week, and it will be the first time I'll be in the water since Doreen's death. I know I'll go in, but I also know I'll think twice about it.

Professor Di Twigg
Friend of Doreen Collyer, who was fatally bitten by a shark on 5 June 2016, Perth, Western Australia

the island in 2007, yet a directed cull of bull and tiger sharks was conducted in 2013.

Since many people can see merit in removing potentially dangerous sharks from areas with high human overlap, and thus support lethal mitigation methods, perhaps the greatest controversy in many of the methods is the unacceptably high percentage of non-target species bycatch. It should be acknowledged that more recent mitigation measures have responded to this criticism, and have begun to modify traditional methods in ways that greatly reduce non-target species bycatch. Some even aim to release potentially dangerous target species alive into areas that are considered to be outside the threat to humans.

Placebo policies and the act of being perceived to take action

While 'placebo policies' do not (supposedly) exist, some legislation or measures seem to be employed for that very function. Stringer and Richardson (1980) coined the term in relation to policies that are visible overreactions by policy-makers, designed to appease the public's fear of certain outcomes that could threaten the popularity of the incumbent government. Placebo policies rely on little more than the general public's wide-scale misunderstanding or lack of knowledge. They are not based on science or even on polled opinion, and may be viewed by some as kneejerk reactions.

Sunstein and Zeckhauser (2008) supported governmental use of 'fear placebo' legislation in response to public demand for excessive action in the face of low-probability risks. They suggested that, in such situations, governments should devise actions that have high visibility at a low cost, yet would have perceived effectiveness. The policies would limit the costs associated with public fear (e.g. reduced tourism, the time government officials would need to spend addressing the issue) and placate irrational public response.

There is a term, *de minimis* (small things), which has been abbreviated from the Latin phrase *de minimis non curat lex*, which means 'the law cares not for small things'. The concept is used by risk regulators to define matters that are 'virtually safe' and so small that they can be treated as if they are zero. Similarly, the US National Institute of Health defines *de minimis* risks as those too small to be of social concern, or too small to justify the use of risk management resources. While the exact quantification of *de minimis* is subjective, numbers such as 1 in 10 000 and 1 in 1 000 000 are used. There are *de minimis* caveats in tax rules set by the US Internal Revenue Service and the EU's business competition law. When applied properly, the identification of *de minimis* risks can help set priorities for regulatory action in socially beneficial ways and encourage economic efficiencies. Statistically, shark attack is a *de minimis* risk. So, from a calculated risk management perspective, should we be concerned? Should we be encouraging governments to fund mitigation measures, and should we support research funding for developing technology to lessen the risk of shark attack? While the statistics suggest that we should not, I still believe that in certain regions the threat of shark attack is not necessarily a trifling matter, and therefore should be addressed. To a reasonable and comparable extent, that is. Despite our ability to consciously qualify risk and quantify incidence rates, we remain irrational when it comes to fear. Even with *de minimis* risks, the uncertainty of the events causes us to err on the side of safety. Furthermore, fear does not only affect an individual; it can be felt far and wide – socially, geographically, politically and financially. Although the US has the greatest annual number of shark bites by far, the bites are mostly minor and do not warrant significant concern. The situation is different in Australia, South Africa, Brazil and Réunion Island, though, where fatality rates and percentages are significantly higher. Thus, it is not surprising that these locations are becoming leaders in innovative and effective shark attack mitigation measures and policies. Parenthetically, they should also be leaders in shark, and shark attack, research and conservation.

Opinions on shark attack mitigation range from that of complete conservation, with full onus on humans to mitigate the situation (even if it means staying out of the water entirely), to the other extreme of proactively and indiscriminately killing any and all sharks that could pose a threat to humans. However, most people have views somewhere in the middle; they desire coexistence where sharks are allowed to remain unharmed in their natural environment and humans are able to enjoy natural waterways for recreational purposes without needing to be continuously alert for sharks. Realistically, while doing very little, we are quite close to this. However, there is room for improvement.

Conclusion

Human–wildlife conflict relates to an extensive group of animals; however, sharks are the most geographically diverse offenders. While governments may feel the need to respond to shark attacks, the development of practical and effective mitigation measures is challenging. Regional environmental conditions, shark movement, human activity and resources must all be considered. In creating legislation and policy, it is important that policy-makers take into consideration public opinion and avoid overreaction. While biological considerations are essential to human–wildlife management, one of the most direct and beneficial responses a government can make is to manage human behaviour and public perception by acting to calm, placate and regulate the public. In some circumstances, this may involve the implementation of placebo policies.

9

Regional shark attack mitigation measures: what are they based on and do they work?

Although often demanded by members of the public in panic situations, or after particularly distressing, repeat or highly publicised incidents, the decision to deploy regional shark attack mitigation measures should not be taken lightly. To act, or not to act, would be only the first of many decisions. Subsequent matters requiring careful deliberation include the environment that the equipment will be deployed in (considering equipment set-up, maintenance and long-term durability), the area of coverage where protection is needed, potential adverse impacts on other flora, fauna or natural structures (including all seasonal animal migrations through the area), the potential for human health implications (entanglement, aircraft malfunctions, conflicts with medical devices such as pacemakers, or the potential for human discomfort caused by the device), stakeholder opinion, cost (initial equipment costs, plus set-up and maintenance costs), the species of sharks in the area (and the tested efficacy of the device against those species), patterns of human water use (recreational and commercial), shark attack history in the area, and the trialled or documented efficacy of the technology. These criteria can be seriously limiting when weighing up potential mitigation methods. However, they need to be comprehensively addressed on a location by location basis, as they would have a significant impact on the overall efficacy of the mitigation result (and on stakeholder satisfaction). Recently, the New South Wales (Australia) Government signed an AU$2.6 million contract with Eco

Shark Barriers and Global Marine Enclosures for the 'design, construction, transportation, installation, maintenance and removal' of two shark barriers on the state's north coast (Dole 2016). The idea certainly had merit, and appeased the local population in providing an environmentally friendly option for safe swimming beaches. However, after attempting to install the barriers, the project was abandoned as local weather and tidal conditions rendered the design inappropriate for the proposed beaches – something that should have been determined much earlier in the process through adequate site assessment and consultation with the designer. The abandoned design resulted in a waste of resources that could have been better allocated to more appropriate measures.

A wide variety of regional shark attack mitigation measures have been used and trialled. One of the most obvious delineations between methods is lethality – some methods are intended to remove sharks from the environment, whereas others are intended to deter sharks. Lethal methods generally include the more traditional efforts at regional mitigation. However, many are being altered or phased out in favour of non-lethal or less lethal measures due to their poor popularity among the general public. Mitigation measures may also be differentiated as active or passive, or those reliant on human monitoring. As with personal devices, many regional mitigation methods rely on the exploitation of shark sensory capabilities, and methods can be divided according to the sensory system targeted. Shark attack mitigation methods, and indeed all human–wildlife conflict mitigation measures, can also, somewhat ironically, be categorised as methods that manage the hazard (e.g. wild animals) or those that manage people, whose actions can be far more unpredictable, destructive and hazardous.

Few methods, including long-standing measures like beach mesh nets, have been through rigorous scientific testing to determine their efficacy in preventing shark–human interaction. And, as with just about everything relating to shark attack, the various mitigation measures have all attracted their own degree of controversy. Their value in relation to mitigating the risk is often highly debated between the stakeholders (technology developers, managers and operators of the

On Sunday the 10th of February 1974, I was on duty as a lifesaver and my neighbour from across the road came up to me and asked, 'Is it safe to swim here?' And I said to her, completely affronted, 'What do you mean is it safe to swim here?', knowing exactly what she meant. Someone had been attacked by a shark the previous month. And I said, 'You'd be more likely to be struck by lightning than be bitten by a shark. It's a once in a lifetime occurrence, and it will never happen here again.' And there were shark nets, so I was absolutely assured that there was no risk from sharks. I was bitten three days later. So the classic 'it will never happen to me' story happened to me. Maybe I don't learn, but I have since been quoted as saying, 'You have a 1 in 7.5 million chance of being bitten every time you go in, so I figure I've got ~7.4 million swims to go before I need to be concerned.' But that's just a joke. I have a certain concern, and it goes through my head, but it has not stopped me from swimming in the ocean.

The attack happened at Inyoni Rocks, Amanzimtoti, 26 km south-west of Durban, South Africa. I was 14 years old at the time, but I remember it vividly, like it happened five minutes ago. I was a junior lifesaver, and we were having our last training session before hosting the provincial state champs on our beach the coming weekend. To accommodate for work/school schedules, we were training into the early evening and the very last thing that we did was a surf swim. We swam out to where the breakers start, which was ~25–30 m out. Then, all of a sudden, a guy we called Jo Kool yelled out, 'Everyone swim for shore' and he started swimming. I followed right behind him. What I didn't know at the time was that a shark had bumped him, and as a result he had a whole lot of gashes on his knee and was bleeding, so I was literally swimming into his blood trail. He got to shore, put his feet down and ran out of the water. I was just about to pull my feet underneath me and run out as well, so the water was probably about a metre deep and I was probably about 2 m from the shore. The shark grabbed my right calf from behind and completely shook me out of the water and then dragged me under. The next thing I knew, I was on the sand with the wave washing back, and there was all this blood. And to this day, I can still feel the muscle in the back of my knee wobble like it did at that time. I was screaming and pushing backwards up the beach and two guys came to help me. They laid me down, and my brother grabbed my towel and applied a tourniquet. They phoned the local doctor and he set up a blood transfer. Then, after 30 minutes, I was loaded into an ambulance, because that was standard shark attack practice at the time.

The funny thing was it didn't hurt. When you stroke a shark from nose to tail, it's quite smooth, but if you go the other way, it's really rough. In the splashing around, the shark grazed my left leg just above my ankle. During treatment, someone raised my left leg, which was largely unaffected, but they had their thumb on that little graze and I had to ask them to move it because it was hurting me. So people find that a little bizarre – there I was, with my calf ripped off and I was missing most of my fibula, and that didn't hurt, but the pressure of a thumb on that very slight graze did. At no stage can I ever say that there was any pain from the bite itself. Even when the shark had me in its teeth, there was no pain. It was just like a huge force, like an iron filing being attracted to a magnet. I can't say it was biting, crushing, tearing, or anything like that. It was just a huge force.

When I came around after surgery, I asked if they had amputated the leg, and they said yes. I tried to sit up to look because I wanted to see, but they kept pushing me down. The next day, I remembered that they had told me that the leg had been amputated, but I could feel my toes. No one had told me about phantom feelings or phantom pains, but sure enough, it was gone. And I can remember going through a thought process, and thinking, 'What am I sup-posed to do? How am I supposed to behave? Do people expect me to cry?' No one teaches a young person what to do when some creature rips their leg off. I figured I could cry if I wanted to. But then I thought, 'If I do cry, what is that going to do and how is that going to help? It's not going to bring my leg back.' So I kind of worked through this process on my own, and thought, 'Okay, if it's gone, it's gone, and nothing can bring it back, so I've just got to get on with life and do what I damn well can.'

I had been doing springboard diving before the attack, but I figured it was pretty obvious that I couldn't do that anymore, so I decided to train as a diving judge. I went to a training course in September and during a lunch break my mum suggested that I just try to dive. I left the 1 m board alone and went straight up onto the 3 m board. The first time I crawled to the end and stood up and did a one and a half somersault off. And I thought, 'Oh, I can do this'. I sub-sequently represented the Western Province in the National Championships for nine years in a row. I also kept swimming, and won the 20 km Geo Bay swim, breaking the record by 20 minutes.

There were shark nets at the beach where I was bitten. But the shark nets were meant to be serviced every day. If the shark mesher wasn't able to get out, he was allowed to leave it to a second day, but if he couldn't get out the second day, his contracted duty was to inform lifesavers, who would put a blanket ban on bathing until the nets could be serviced. We'd had rough seas and floods and the mesher hadn't been out for three weeks, but he hadn't let the lifesavers know. When the nets did come out after my attack, they were absolutely tan-gled with debris and trees and branches and things, so they would have been completely ineffective. I do wonder if the shark would have still been there even if the nets had been functioning properly. Overall, there were five attacks on six people within a period of 13 months. In three cases, the nets were there and intact; for me and Jo, the nets weren't intact. It was only the last guy who was in a location without nets. So the nets weren't really doing much. While I have no direct evidence of this, the reports were that most of the sharks caught in the nets were caught going out. So they had been in, on the beach side, and then got caught heading back out to sea. The nets where we were weren't as big as the ones in Durban. They ended 10 feet above the bottom and they started about 5 or 6 feet below the surface. And they were sort of staggered, whereas in

Durban there was one continuous net that went from surface to seafloor. There were never any attacks on the Durban beaches during that time, so those nets are seemingly a lot more effective.

They were only able to determine that the shark that bit me was a dusky or Zambesi shark [bull shark]. Great whites were actively hunted for trophies in South Africa at the time and there was a population explosion of duskies and Zambesis. I don't have any scientific evidence of this, but having studied zoology, I think that the apex predator had been taken out and that had caused the population explosion of the other sharks. And for that reason, I'm against culling and drumlining, and actively targeting great white sharks. I'm more in favour of risk avoidance measures, like the apps that track sharks. Information is key. I've considered personal mitigation devices and actually tried to buy a Shark Shield, but the website wouldn't work. Then there was that kid in America who was bitten wearing a Sharkbanz, and so I thought, 'Hmmm, maybe I won't bother spending the money.' I'm entered in a 5 km ocean swim in February at Manly, and they hang one of those electronic Shark Shields at every buoy, so that makes me feel better about the swim.

About a year after I was bitten, the movie *Jaws* came out. A newspaper reporter took me and one of the other guys who was bitten to see it, and we watched it. I think both of us had the same sort of reaction; for us, it just wasn't realistic. It was overplayed and overdramatised, so it didn't really affect either of us.

Damon Kendrick

Full details of the incident, including a thorough description of the event, the environmental conditions during the day and at the time of the attack, a description of the injury, the first aid and medical treatment received and a description of the attempts made to identify the species can be found at http://swimhistory.org/champions/international-stars/damon-kendrick. A video of Damon doing the Rottnest Channel Swim can be found at https://vimeo.com/72464942.

equipment, legislative bodies, researchers, the public, environmental groups). Simply put, this means that effective shark attack mitigation is complex and unsuited to a 'one solution fits all' approach. Excluding a global cull of all sharks, a uniform solution to shark attack mitigation is impossible. However, I fully believe that the development of a multifaceted approach that will successfully mitigate a significant percentage of shark attacks is entirely possible in the near future.

Lethal mitigation measures

Long-term lethal mitigation measures have been continuously employed in South Africa, New South Wales and Queensland (Australia). Lethal measures have transiently or recently been employed in Hawaii, Dunedin (New Zealand), Recife, Egypt, Russia, the Seychelles, Mexico, Réunion Island and other locations within Australia. The demand for lethal mitigation action seems to be most relevant in popular coastal locations that rely heavily on marine-based tourism and revenue (Gibbs and Warren 2015). However, the efficacy of such programs is yet to be proven (Wetherbee *et al.* 1994), and the consequences to shark populations have been enormous. For example, the New South Wales beach nets caught 16 064 animals between 1950 and 2008, including 12 354 sharks (DPI 2009), and a report detailing the 2014–2015 meshing season revealed the entanglement of 189 animals (Daly and Peddemors 2015). It has been suggested that the negative outcomes for sharks are afforded lower political priority than human social and political outcomes, and the improvement of public perception is more important than the means by which this is accomplished (Neff 2012). Often, the perception of risk reduction is considered more important than actual risk reduction.

Culling and the purposeful removal of animals

Culling

Culling is the active removal of targeted animals from the population, and may be employed for two different reasons. The first is a more conservative, but reactive, approach, which aims to track and kill animals that have attacked humans or been implicated in near-misses. The logic behind this type of culling is that if a shark has bitten once, it may have acquired positive reinforcement (food) from that action and thus may bite a human again. The concept of repeated attacks on humans by an individual shark ('rogue sharks') has been neither confirmed nor disproven by research. Evidence from white sharks (*Carcharodon carcharias*) suggests that this sort of behaviour would be unlikely (Neff and Hueter 2013). Yet an individual oceanic whitetip shark (*Carcharhinus longimanus*) bit three humans over five days (including one fatal bite) in Egyptian waters; that shark had previously been hand-fed (conditioned) by a diver (Levine *et al.* 2014). Directed culling following an attack or an attempted attack has been employed in Western Australia on a special-order basis since 2001. However, most efforts to catch offending animals have failed, while others have resulted in the destruction of multiple animals. The success of such a program is highly correlated with the timing of the order and the rapidity of the response (which may be delayed due to weather, sea conditions or crew availability). It is also extremely difficult to confirm that a caught shark was in fact the offending animal before (and in many cases, even after) killing it.

The second form of culling is a more passive, less conservative, proactive approach, with the aim of reducing overall shark numbers. The idea of proactive culls is based on the premise that increasing shark attack incidence is a result of shark overabundance. There are some notable assumptions in this reasoning. The obvious first assumption is that shark populations are overabundant, a topic that is often discussed anecdotally. Although some regional populations have been shown to be increasing, most research suggests that the majority of shark populations are stable or declining. And this has been the case for decades. Even if regional populations are increasing, they are still likely to be low in comparison to natural/historical levels and species carrying capacities. Although seemingly logical, the premise that more sharks lead to more attacks is unproven: in fact, studies have shown that shark population numbers are not linked to attack rates on prey (Brown *et al.* 2010). Another assumption is that sharks target humans as prey. This is certainly not proven and in fact the contrary is often argued; that is, many believe that sharks do not target humans as we are not a prey item that featured in their successful evolutionary history. An analysis of shark attack files in the early 1970s found that 69% of negative encounters (with a variety of species) resulted in only one bite, and an additional 15% involved two bites (but possibly from within an individual bite sequence; Baldridge 1974). This behaviour is quite inconsistent with the behaviour of a predator actively and purposefully trying to consume its prey. Many attacks on humans are non-fatal, and even fatal attacks often do not result in consumption. However, this argument is difficult to defend in the rare cases where the body is never found or where repeated bites are sustained.

Another argument for the ineffectiveness of culling is that many large, potentially dangerous species of sharks are migratory and are therefore only transient visitors to various locations. They often utilise different regions during different stages of their life, or in response to various environmental, reproductive or predatory drivers. For example, tiger sharks (*Galeocerdo cuvier*) are known to display wide-ranging movements even within their home ranges. Their visits to coastal sites are often brief and infrequent, suggesting that culling programs for this species would likely be ineffective for catching a shark implicated in an attack or for reducing the risk of future attacks (Meyer *et al.* 2009a). Thus, for culls to effectively work as a mitigation method, they would have to be continuous and rigorous. While possibly protecting an area from that particular threat, such a degree of action would have serious consequences for the species and the local ecology.

Between 1959 and 1976, Hawaii initiated six shark fishing (culling) programs aimed at reducing shark numbers in the area. A total of 4668 sharks were killed during this time, at an average cost of US$182 per shark (in the currency of the time; Wetherbee *et al.* 1994). Each program was deemed successful, as the catch rate of sharks declined over the course of individual programs. However, factors such as seasonality, annual population fluctuations, weather and discrepancies in fishing techniques were not accounted for. And while shark numbers during the programs were definitely reduced (potentially by 50–90%), the effectiveness of the programs in terms of human safety are not apparent. The number of annual attacks in the region during the 18 years of the culling programs was identical to that of the 18 years before the programs commenced. While attack prevalence in Hawaii has increased since the culling programs ended, it has done so at a rate that more likely mirrors the increased reporting of cases and the highly significant increases in human population and tourism.

Following an increase in acute shark attack incidence at Réunion Island, a cull of 90 sharks (45 bull sharks, *Carcharhinus leucas*, and 45 tiger sharks) was implemented in 2013. The effect of that cull in relation to shark attack mitigation has not been rigorously assessed, but attacks off the island have continued.

There have been 23 shark-related fatalities in Western Australia in the last 100 years, with the most recent 11 fatalities occurring in the last 10 years. Since 2008, the Liberal National Government has invested more than AU$33 million in a broad range of shark hazard mitigation measures including aerial and beach patrols, new jet skis and watchtowers, research into shark behaviour, shark population estimates and non-lethal detection and deterrent technologies, satellite buoys monitoring tagged shark movements off the Western Australian coastline, sending real-time alerts to beach managers and the public, a SharkSmart website (http://www.sharksmart.com.au), beach enclosures and a recently announced three-month trial of drone technology. Understandably though, this unprecedented concentration of fatalities continues to have a significant impact on people's perceptions of the ocean and their enjoyment of water-based activities.

Potentially dangerous sharks in close proximity to populated areas represent a serious threat to public safety, and under certain conditions I support the fishing for these sharks. Strict criteria have been developed to assist decision-makers in implementing the serious threat policy which allows for capture gear to be deployed:

- when a person has been seriously injured or killed by a shark, or involved in a shark-related incident that could have resulted in a fatality; or
- when there have been confirmed reports of what could reasonably be considered to be the same shark or sharks, 3 m or more in length, over three days in the same or similar location, within 1 km of the coast, where people are likely to be using the water, and where other actions to ensure public safety have been, or are likely to be, ineffective.

Further information on the strategy, including the legislative context in Western Australia, is available in the Public Environmental Review, the Western Australian Shark Hazard Mitigation Drum Line Program Review 2013–14, and the government's overall shark hazard mitigation strategy, all of which are available for download from http://www.dpc.wa.qov.au.

Shark hazard management is a complex, emotive and divisive issue, with no definitive solution. Measures implemented in isolation will not be effective, but a comprehensive strategy including a broad range of complementary measures will reduce the risk of human–shark interactions.

The Liberal National Government is committed to addressing its duty of care to minimise the risk of shark attack by supporting existing measures, investigating ways to enhance and expand these measures and researching new approaches as they become available.

Colin Barnett MLA
Premier of Western Australia

Shark meshing programs

Shark meshing programs, also known as beach netting programs, are probably one of the more traditional and well-known mitigation methods. However, their structure and design are often misunderstood, their effectiveness is overestimated and their use has become quite controversial (Gibbs and Warren 2015). Shark mesh netting was introduced as a way to catch and kill potentially dangerous sharks, in New South Wales in 1937. Similar measures were employed in Durban in 1952, with subsequent extension along the Natal coast, and in Queensland in 1962. Although other regions trial these types of nets on occasions, the three original locations are the only established, long-standing shark netting programs. The Queensland program is by far the largest, covering 85 beaches over ~2000 km of coastline. This is followed by the New South Wales program, which covers 51 beaches over 200 km of coastline, and the KwaZulu-Natal program, which covers 37 beaches over roughly 300 km. Shorter mesh netting programs have been undertaken in Brazil, New Zealand and Hawaii. Shark nets are generally promoted by local governments as increasing beach user safety and boosting tourism in protected areas.

The design, concept, usage and functionality of shark mesh nets are often poorly understood by the public. Nets in no way form a barrier – around 40% of sharks are caught on the beach side of the net (Dixon 2002; McPhee 2012) – so it is not much more than chance as to whether a shark will be on the 'right' side or the 'wrong' side of the net. The nets are designed to function as population control measures, through the capture and killing of animals. Nets are not baited: they are a passive method of animal collection, relying on movement paths taking animals into the proximity of the nets. However, it could be argued that other animals caught in the net, in distress and trying to free themselves, or dead and decomposing, could serve as attractants (bait) to sharks in the vicinity. Nets in the three major netting programs use a stretched mesh size of 50–60 cm. They stretch over a finite distance, and do not isolate entire beaches. Net length ranges from 150 m in New South Wales, to 186 m in Queensland and 214 m in KwaZulu-Natal (where nets are also set in parallel pairs). Nets ~6 m high are generally set in 10–14 m of water, and can float or be sunken. Thus, they do not extend to both (or either) the substrate and the water surface and cover approximately half of the water column. Nets are generally set parallel to the shoreline, in consideration of tidal and current patterns. Their distance from the shore is not fixed, but is determined by local topographical attributes and sea conditions. In most cases, they are deployed around 400–500 m off-shore. Mesh nets effectively entangle animals, trapping them in place. While the entanglement and trauma from struggling against the net can kill animals, drowning would be the most common cause of death. Most species of sharks (in particular, the larger, more mobile species) are obligate ram ventilators, meaning that they must be able to swim forward to push oxygenated water over their gills for respiration. If unable to do so, these sharks drown. Similarly, trapped air-breathing marine mammals and reptiles would not be able to get to the surface for oxygen. It is largely the variation in respiratory requirements that causes certain species to be more susceptible to net fatality (e.g. rays that do not need to move to breathe can typically be released alive). This is also the reason why nets checked more regularly have higher live animal release rates. The nets in the three original programs are checked at differing intervals: daily (Monday to Friday) at KwaZulu-Natal, every two days in Queensland and every three days in New South Wales, although all checks are weather permitting and reliant on the work ethic of the contractors. In KwaZulu-Natal, non-functional nets (due to entanglement or displacement) result in total local beach closure. Nets are not always deployed on a daily basis: they may be removed seasonally or temporarily due to poor weather/environmental conditions. Nets are removed from the Kwa-Zulu Natal beaches annually during the sardine run due to the extreme amount of net mortality to targeted and non-targeted sharks, marine mammals and birds during that time. Similarly, the New South Wales netting program runs for only eight months of the year (September to April) to avoid the peak whale migration season.

The three major shark mesh netting programs collectively catch up to 2500 sharks per year (Dudley and Gribble 1999). While the effectiveness of nets in terms of shark attack mitigation is generally considered equivocal, the mortality caused by the nets has been shown to have a major negative effect on both local and migratory shark populations (Dudley and Cliff 1993; Dudley 1997). Indeed, the New South Wales shark mesh netting program is listed as a key threatening process in the *Fisheries Management Act 1994* and the *Threatened Species Conservation Act 1995* (NSW). Key threatening processes are defined by the New South Wales *Fisheries Management Act 1994* as 'a process that threatens, or that may threaten, the survival or evolutionary development of species, populations or ecological communities of fish or marine vegetation'. Potentially dangerous sharks are not the only casualties of beach mesh nets, and this indiscriminate killing of both target and non-target marine animals (including many threatened and/ or protected species) has made the method highly controversial. The controversy is partly due to a greater understanding of the low natural population numbers of many species of sharks, their conservation status, the role of sharks in the marine ecosystem and the acknowledgment of the staggering anthropomorphic mortality of sharks from a variety of sources (Simpfendorfer *et al.* 2011; O'Connell and de Jonge 2014; Gibbs and Warren 2015). Despite attempts to reduce the amount of non-target bycatch in shark mesh nets, an assortment of animals are still caught regularly, including marine mammals, marine turtles, teleost fish and non-target sharks and rays.

Openly accessible data from the Queensland Shark Control Program (which includes drumlines and mesh nets) shows that the program removed 42 white, 1248 tiger and 445 bull sharks from the population between 2011 and 2015. The government considers the program to be successful. However, at least 31 species of sharks were caught, only five of which were implicated in bites on humans in Australian waters over the same period. Despite the program employing nets specifically designed to catch sharks greater than 2 m in length, less than half (43%) of the sharks caught were above that length and a decent proportion of the sharks over 2 m were non-dangerous species. Although the data provided by the program were somewhat inconsistent for 2015, that was the first year that catches were individually quantified to detail the catch method (net versus drumline). From the details provided, beach mesh nets collected only 11% of the total catch. White, tiger and bull sharks were caught by both methods, but the nets were responsible for only 36%, 3% and 22% of species catches, respectively. Along with the 695 sharks caught by the Queensland Shark Patrol Program in 2015 (71 by nets, 556 by drumlines and 68 undocumented), 82 other marine animals were also caught (46 in the nets and 36 on the drumlines). While the shark catch rate (although not necessarily target species-specific) for the drumlines does appear to show quite a high degree of selectivity and effectiveness for sharks, the net catch equates to roughly one non-target animal in the nets for every two sharks (which may or may not be potentially dangerous species). The New South Wales program reported a total of 189 animals entangled in the nets in the eight-month 2014/15 season. Of those, 23 (12%) were threatened or protected species (including targeted white sharks), and 131 (69%) were non-target species. The mortality rate for both target and non-target animals was incredibly high for all species, with the exception of rays.

According to operators, shark mesh netting programs boast high bather protection success rates. Only one serious (non-fatal) attack has occurred on a bather at netted New South Wales beaches in recent times. The KwaZulu-Natal Shark Board details only four serious (non-fatal) attacks at netted beaches under its jurisdiction in the three decades following the implementation of nets. Statistics reported by Cliff (1991), however, state that 15 of the 46 (33%) attacks in Natal between 1960 and 1990 occurred at, or in close proximity to, beaches protected by nets. The reported success of the programs must also acknowledge contributions from other mitigation measures in place. Queensland has always employed a combination of mesh nets and baited drumlines, New South Wales employs aerial surveillance and KwaZulu-Natal benefits from the Shark Spotters program. Many netted beaches are also patrolled beaches. While it's not possible to directly link the effectiveness of shark mesh nets to a reduction in serious shark bites, it is argued that the number of sharks in an area (which nets are designed to target) does not correlate with shark attack statistics on humans. A lack of correlation between shark numbers and attacks on more natural prey has been shown with northern elephant seals (*Mirounga angustirostris*) and white sharks (Brown *et al.* 2010).

An often overlooked, but destructive, side effect of shark mesh netting programs is the potential for lost or displaced nets. If not recovered, they would add to the hundreds of kilometres of nets that annually accumulate in natural waterways ('ghost nets') and continue to collect and kill a wide range of aquatic animals, as well as destroying aquatic environments and benthic structures. In April 2015 alone, five nets were reported missing from the New South Wales shark netting program following severe storm activity (Daly and Peddemors 2015).

Drumlines

Drumlines have been employed in a few regions. Their design and purpose, however, have been reasonably varied. The traditional use of drumlines as a shark mitigation method works on the same premise as mesh nets. The idea is to decrease the local and transient population of large, potentially dangerous sharks in a particular area to reduce the likelihood of shark–human interaction on beaches in close proximity. Drumlines are a passive, autonomous and often fatal method of shark population control, which relies on an animal being attracted to the bait and becoming hooked. Although drumlines catch a variety of non-target species, the percentage of bycatch is greatly reduced through this method compared to that of mesh nets.

Drumline hooks are suspended from large plastic floats via a set distance of trace and lead (in Queensland, 2 m leads with adjustable trace to account for water depth) and are anchored to the seabed with a combination of rope and chain to prevent them being carried away by current or tide. Hook size and shape depend on the purpose of the program (catch and kill or catch and release). The soak time (time between the lines being set and checked) also varies between programs (e.g. every second day in Queensland, daily in South Africa).

The dominant causes of mortality in animals hooked by drumlines, depending on line length, are either drowning or exhaustion due to the prevention of the swim–glide pattern used by various shark species for energy conservation. However, more recent drumline mitigation programs have included measures to reduce catch mortality. As such, drumlines could realistically become an effective yet controlled method for permanently removing or displacing large and potentially dangerous

The Queensland Government's Shark Control Program (the Program) commenced in 1962. The purpose of the Program is to reduce the risk of shark attacks on people enjoying Queensland's more popular beaches. Only one fatality has occurred on a protected beach since the Program commenced despite a large increase in the number of people swimming at those beaches over the same period.

The Department of Agriculture and Fisheries (DAF) is responsible for administering the *Fisheries Act 1994* (the Act). One of the purposes of the Act is to reduce the possibility of shark attacks on humans in coastal waters of the state adjacent to coastal beaches used for swimming. The Queensland Government achieves this purpose through the delivery of the Program.

The Program has invested significant resources into researching alternative shark control methods, including shark-detecting sonar systems. This technology is still developmental and, as a consequence, traditional control methods remain the most effective to reduce the risk of shark attacks.

The Program has evolved over the 54 years of its operation, incorporating new technology and methods as information and knowledge increase. It continues to trial and has used various types of acoustic alarms, or pingers, over many years in an attempt to minimise impacts on non-target species. In 2014, nets were fitted with a new type of acoustic pinger which has resulted in a 40% reduction in the number of dolphins caught. The ultimate success of this pinger remains to be seen. These early results are very encouraging.

DAF continues to monitor the progress of trials being conducted in New South Wales. If new technologies are shown to be effective in preventing marine life fatalities and are practical for use, they would be considered as part of the Queensland Program.

The Program also invests significant resources in educating swimmers on how to minimise the chance of a shark attack. These tips include:

- swim or surf only at patrolled beaches – between the flags and where shark safety equipment is in place;
- obey lifesavers' and lifeguards' advice, and heed all sign and safety warnings;
- leave the water immediately if a shark is sighted;
- do not swim or surf after dusk, at night, or before dawn when sharks become more active;
- do not swim or surf in murky or silt-laden waters;
- do not swim in or at the mouth of rivers, estuaries, artificial canals and lakes;
- never swim alone;
- never swim when bleeding;
- do not swim near schools of fish or where fish are being cleaned;
- do not swim near, or interfere with, shark control equipment;
- do not swim with animals.

Mark Biddulph
Deputy Chief of Staff, Parliament and Policy, Queensland Government

sharks, while minimising non-target bycatch. For these alterations to limit mortality, though, it is vital to first understand what contributes to the lethality of the drumline apparatus, and how the animals (target and bycatch) respond physiologically over both the short and long term.

Drumline mortalities are not limited to the easily identified and quantifiable capture mortalities (i.e. animals that are dead or terminal upon collection); 'hidden' post-release mortalities also need to be considered. The quantification of post-release mortality is far more complex, due to the unpredictability of injuries, handling, animal stress levels, the species' physiology and the environmental conditions at the time of release. Thus, most shark control programs completely disregard the potential for post-release mortality, and instead imply that all animals released alive will survive. It has been suggested that post-release mortality could prove a greater source of mortality than harvest mortality (Cramer 2004; Douglas *et al.* 2010; Molina and Cooke 2012). The compounding effects of physiological exhaustion from fighting against drumlines would be a cause of rapid or latent mortality, and catch and release survivorship data have shown that latent mortality can be seen up to 48 days post-release (Campana *et al.* 2016). Trauma, hypoxia (low blood oxygen levels) and exercise-induced stress can contribute to rapid mortality following release (Renshaw *et al.* 2012; Skomal and Mandelman 2012), while reduced feeding ability (from hook trauma to the mouth or jaw) could result in more prolonged mortality. Furthermore, while not directly lethal, potential effects of capture, such as altered behaviour, reduced growth or disruptions to reproductive activity, could lead to reduced individual fitness (Wilson *et al.* 2014).

Animals injured through the hooking, collection or release process have significantly higher post-release mortality rates (Campana *et al.* 2016), although species-specific activity levels, physiology and environmental conditions will affect an individual's response and capacity to survive hooking, handling and release. Gallagher *et al.* (2014) performed the most comprehensive study to date on post-catch stress, reflex impairment and survival rate of five species of sharks that most closely match the species caught on shark control program drumlines. The study population consisted of subadult and adult blacktip sharks (*Carcharhinus limbatus*), bull sharks, great hammerhead sharks (*Sphyrna mokarran*), lemon sharks (*Negaprion brevirostris*) and tiger sharks. Sharks were caught using drumlines fitted with circle hooks and enough line to allow the sharks to swim in a 23 m radius. The lines were checked after an hour of soak time. This methodology notably contrasts with the much shorter line used in shark control programs, and the far greater soak time. It was found that the five species showed significant differences in post-catch blood chemistry, reflex impairment and four-week post-release survival. Tiger sharks were by far the most hardy and resistant to catch stress and mortality in all accounts, followed by lemon, bull, blacktip and great hammerhead sharks. While tiger sharks had essentially 100% survival at four weeks post-release, bull sharks showed 74.1% survival and great hammerhead survival was just 53.6%. Species that display a more limited aerobic capacity, such as blacktip and dusky sharks (*Carcharhinus obscurus*), are associated with higher catch and post-release mortality rates, compared to less active species such as tiger sharks (Marshall *et al.* 2012; Gallagher *et al.* 2014). Factors that increase the likelihood of post-release mortality in hooked sharks include increased hook soak time, the use of J hooks rather than circle hooks, warmer water temperatures (potentially due to decreased dissolved oxygen levels in warmer water and/or the more rapid accumulation of blood lactate in warmer water temperatures) (Danylchuk *et al.* 2014; French *et al.* 2015), smaller individual size, and the practices used by handlers (Diaz and Serafy 2005; Campana *et al.* 2009; Godin *et al.* 2012; Danylchuk *et al.* 2014; Gallagher *et al.* 2014). Hook selection is critical in ensuring the most discriminate catch, when the goal is non-lethality or selective lethality. The two main hook designs used on drumlines are J hooks (Queensland and South Africa) and circle hooks (Recife). Circle hooks are designed to hook into the animal's jaw, whereas the likelihood of foul hooking (hooking in areas other than the mouth) is far greater with J hooks. Foul hooking locations include the gills, throat and gut, among others, and this significantly increases the probability of rapid or long-term animal mortality through excessive blood loss (generally through gill hooking or severe internal trauma) and unretrieved internalised hooks. Internalised hooks can easily become lodged in the pericardium or other vital organs, or cause chronic systemic disease (Adams *et al.* 2015).

A new wave of 'traditional' technology

Increased levels of environmental responsibility and accountability and increased pressure from the general public and environmental groups have resulted in the need for long-standing and indiscriminate lethal programs (using what could be considered antiquated technology), such as mesh nets and drumlines, to be reconsidered and reworked. Many governments have accepted this challenge and are phasing out old technology in favour of new, more environmentally friendly and ethical equipment or adaptations.

SMART (Shark Management Alert in Real Time) drumlines incorporate technology that sends an alert to a response team when an animal is hooked. Responders can immediately attend to the animal and carry out the desired procedure, for example release of non-target animals or relocation/destruction of target animals. The drumlines have been trialled at Réunion Island and are in regular use in certain locations along the New South Wales coast. SMART drumlines, however, involve higher costs and significant resources are required to have a response team on standby when the drumlines are deployed (thus also limiting deployment to when a response team is available and on-hand). Another limitation is the fact that an immediate response would be weather- and condition-dependent. While this technology is a more conservation-friendly option, we still need a better understanding of species' post-release movements if potentially dangerous animals are to be relocated and released. Juvenile tiger sharks have been shown to move to deeper, off-shore water (>100 m) immediately after capture and tagging for approximately two weeks, but then return to a shallow distribution. However, larger tiger sharks did not necessarily show the same alteration in activity (Hazin *et al.* 2013). Juvenile lemon sharks (*Negaprion brevirostris*) and Port Jackson sharks (*Heterodontus portusjacksoni*) are capable of homing, or returning to their spatially limited home ranges, when transported 16 km and 3 km, respectively (O'Gower 1995; Edrén and Gruber 2005). Therefore, before relying too heavily on conservation-friendly relocation programs in shark attack mitigation measures, extensive multi-species tracking studies of potentially dangerous relocated animals are necessary. However, SMART drumlines are a promising measure in terms of minimising bycatch and non-target shark mortality.

Mesh net technology has remained relatively static, but the operational procedures for its use can have a significant effect on its environmental footprint. Checking the nets on a daily basis, instead of every couple or every few days, provides a much higher live animal release rate. Similarly, the removal of nets during annual whale or fish migration seasons, as is done in New South Wales and KwaZulu-Natal, limits the amount of bycatch collected in those programs. However, it obviously also removes any protective capacity afforded by the nets.

Recent funding has been allocated for the development of acoustic systems that are said to act as virtual shark nets, and may be able to provide an environmentally friendly replacement for nets.

Non-lethal regional measures

Although a proportion of people hold more extreme views on shark attack mitigation measures, and prefer the removal of potentially dangerous sharks from the environment entirely, most people prefer measures that facilitate coexistence between sharks and humans in shared waterways. Several different measures have been trialled and employed for this purpose.

Public bans

Various regions have implemented temporal, seasonal or permanent bans on particular beaches or recreational water activities in response to the threat of shark attack. The state government of Pernambuco (Brazil) introduced a seasonal ban on surfing (and related activities) at certain beaches in May 1999, and Réunion Island prohibited swimming, surfing and body-boarding in particular areas in 2013. More commonly, individual beaches are temporarily closed following a shark attack or shark sighting. While effective (if people are not in the water, they cannot be bitten by a shark), the bans are highly unpopular and

Shark management

The NSW Government announced a new five-year, AU$16 million NSW Shark Management Strategy in October 2015.

Its key objective is to increase protection for swimmers and surfers from shark interactions while minimising harm to marine animals. Some of the emerging technologies that have been trialled under the Strategy include:

- increased aerial surveillance;
- drone trials;
- sonar 'Clever Buoy' technology;
- SMART drumlines which allow sharks to be tagged, relocated and released;
- coastal VR4G listening stations, which provide real-time data of tagged shark movements to beachgoers through the NSW Shark Smart app.

Shark barriers are no longer being trialled at Lennox Head and Ballina after the contractors advised they could not be installed safely and effectively.

In addition, in October 2016, the NSW Government announced a six-month trial of shark nets on North Coast beaches in response to the area's recent increase in shark incidents.

The design and implementation of the trial, including locations, has been guided by advice from the North Coast community.

Support for projects

The NSW Government is offering a total of $200 000 to support national or international projects that are aligned to the NSW Shark Management Strategy, and can increase shark protection for beachgoers when hitting the water.

Key areas for funding aligned with the Strategy include:

- personal shark deterrents;
- area-based shark deterrents;
- shark detection methods;
- shark biology relevant to interactions with humans;
- socio-economics of shark–human interactions.

For more information, please visit: http://www.dpi.nsw.gov.au/fishing/sharks/shark-management/annual-competitive-grants-program.

The NSW Government is conscious that no single program, or combination of open-water programs, can ever totally eliminate shark interactions with the community. With that in mind, this government has made a significant commitment towards understanding and mitigating the risks while minimising harm to marine life.

Dr Geoff Allan
Deputy Director General, DPI Fisheries

A sign warning of the heightened risk of shark attack at a Recife beach. Photo by Fabio Hazin.

regularly disregarded. Although useful over the short term to avoid more risky situations when potentially dangerous animals are known to be in the immediate vicinity, this method does not serve the purpose of allowing humans to benefit from aquatic recreational activities and may decrease the preparedness factor of response teams, which would be problematic if someone disregarded the ban, entered the water and was bitten.

Exclusion nets

Although relatively small-scale in terms of regional mitigation measures, exclusion nets are a highly effective option for some water users. Exclusion nets serve to physically separate sharks and humans. They do not aim to capture or kill sharks or any other marine life, making them much less controversial as well as highly successful. Exclusion nets wholly enclose an area so that sharks cannot penetrate. While highly effective when in good condition, the nets require regular checks and maintenance to ensure their functionality. Their use, however, is limited to locations that can be physically excluded (they can't cover entire regions, or great distances from shore). Being much thicker and more robust than mesh nets, exclusion nets are far more susceptible to environmental conditions and biofouling and therefore must be regularly checked and maintained. These factors limit their protective capacity to water users in calmer environments (i.e. not surf beaches). There is no technical reason why large-scale exclusion nets could not be extended to protect surf beaches; however, the costs for manufacture, deployment and maintenance would be much greater. And although fauna entanglements are greatly minimised by the composition of these nets, they are still capable of catching marine animals, so entanglement rates should be monitored.

An Eco Shark Barrier Net (exclusion net) was installed at Coogee Beach (Western Australia) to form a completely enclosed swimming area. The net is constructed from strong, flexible polymer nylon (as in zip cable ties) with an optional 6 mm plasma rope addition that can be threaded through the net at 300 mm intervals. This provides the net with a 3.5 tonne breaking strain at each rope placement. Floats keep the net at the surface, chains and anchors attach it to the seabed. The net is designed to last up to 10 years in marine conditions and is fully recyclable at the end of its life (http://www.ecosharkbarrier.com.au/the-product/). To date, no animal entanglements have occurred, and fish appear to aggregate around the artificial structure. People local to the area state that the exclusion net is an ideal protective measure for swimmers and bathers and is highly regarded. Water users are happy to drive great distances to swim at Coogee Beach because of the net, despite there being beaches much closer to their homes.

Monitoring

Monitoring programs have the potential to work over a coastal scale or at a regional or beach level, depending on the method. Although regional in coverage, the way that monitoring data are used is up to individuals. The purpose of monitoring programs, in relation to shark attack, is generally to alert water users to the detected presence of a potentially dangerous shark in an area, so that they can make a more informed decision about their water usage. Monitoring programs are non-lethal, proactive and provide a wealth of information that can be utilised to help our understanding of shark presence and movement patterns and trends over time.

Land-based observers

The best-known, tailored, land-based shark monitoring program is the Shark Spotters program developed and employed in Cape Town. The program, founded in 2004, was designed to be a middle-ground operation, supporting both human beach user safety and white shark conservation through a reduction in the spatial overlap of people and sharks. Since its conception, the program has proven its worth and now employs 15–20 spotters over nine beaches (http://sharkspotters.org.za/welcome). Program spotters utilise elevated platforms (e.g. on cliffs or buildings) and scan the surface of coastal waters for the presence of sharks. If sharks are sighted, established protocols are enacted to warn water users and assist people out of the water, if necessary. A flag system, which notifies users of good or bad sighting conditions, high shark alert conditions and whether a shark has recently been spotted, is also employed on the beaches to communicate with and inform beach users. The program is in operation during daylight hours, seven days a week. Observer data are recorded on a daily basis, including environmental conditions, sharks spotted and the number of beach users.

As with any human monitoring program, the Shark Spotters program is susceptible to the limitation of human error and bites have occurred on spotted beaches. Also, environmental conditions that contribute to poor water visibility (e.g. turbidity, wind, glare, darkness) can affect spotting ability. The Cape Town Shark Spotters program has the natural advantage of elevated ground for lookouts and optimal environmental and beach conditions for sighting sharks, including sloping beaches and extended areas of shallow water, which allow sharks to be spotted further from shore. These necessary geographical elements limit the scope of this sort of program at many other locations. A Shark Spotters-type program was the primary recommendation for beaches in northern New South Wales in a government-commissioned study on the best shark attack mitigation methods for immediate investigation and deployment (Cardno 2015), and scoping studies were carried out to assess the likely effectiveness. The results of those studies are not publicly available, but it is known that the program was rejected. Media reports stated that the local council declined to even receive the report about the study, despite purported suggestions that multiple beaches might be suitable for such a program (Coyne 2016).

While lifeguard and lifesaver services generally do not operate from significantly elevated positions, this form of human monitoring is beneficial in mitigating the possibility of shark attack as lifeguards on beach towers just 3 m high have observed sharks up to 300 m off-shore (Parrish and Goto 1997). The ability of lifeguards and lifesavers to respond quickly to the detection of a shark (to clear the water), or to assist and provide advanced first aid to a shark attack victim, would significantly reduce the number of fatal shark attacks at monitored beaches. Not only would medical attention be immediately available, but so would a supply of highly advantageous appropriate first aid materials, and advanced communication capability with paramedics and emergency responders. In the situation of a severe shark bite, this could mean the difference between life and death. Lifesavers are simultaneously beneficial in mitigating a variety of much more statistically relevant beach hazards, like drowning. Thus, the enhancement of lifesaving programs would contribute to reducing the overall number of fatalities at beaches, regardless of the specific risk. The obvious limitation with this sort of measure, however, is the small scale over which it would be effective. Humans can see only so far, even under perfect conditions and from an elevated position, and the ability to see into the water would be reduced from such a close distance. So, while beneficial in many ways, the effectiveness of lifesavers in preventing shark bites (in total isolation from other measures) may be minimal.

Although not a mitigation method *per se*, the need for further research and analysis of shark attack patterns, as well as research into basic shark biology, physiology and movement, is warranted to assist in mitigating shark attacks. The more we understand about trends in shark behaviour (including those directly related to shark–human interaction), the more we will be able to predict high-risk situations. Such information could be fundamental in quickly and effectively stemming the number of shark attacks. This is particularly true given the premise that the primary reason for the global rise in shark attacks is human population increase. Thus, if we can limit the number of people who enter the water

Statement from Surf Life Saving Queensland

SLSQ was invited to provide comment and feedback in relation to a Senate Environment and Communications References Committee investigating shark control strategies currently in place. SLSQ provided feedback across a range of areas as well as some recommendations moving forward to increase protection for beachgoers along Queensland's coastline.

As the state's peak authority on coastal and aquatic safety, SLSQ continues to play a key role when it comes to shark management and prevention, working closely with key stakeholders at all levels to deliver safer beaches and educate swimmers about how to protect themselves in the water.

At a government level, SLSQ works in cooperation and consultation with the Department of Agriculture and Fisheries, and is an active member of the Shark Marine Advisory Group on the Gold Coast. This sees key organisational representatives have direct input into the long-term strategy and day-to-day operations of the Shark Control Program (SCP) in Queensland.

On the ground, SLSQ's lifeguards and lifesavers play an active role in shark prevention and management, particularly as a front-line defence when it comes to monitoring beaches and responding accordingly in the event of a sighting or attack.

By way of background, SLSQ has state-wide standard operating procedures in place for shark sightings, including guidelines that cover when surf lifesavers will act to clear the water and close a beach. Among other things, these state that surf lifesavers will close a beach for at least 60 minutes after a confirmed shark sighting, or until the threat has otherwise subsided. If and when this occurs, lifesavers on duty liaise closely with beachgoers to communicate these processes and ensure that swimmers are kept out of harm's way.

Lifesavers also work proactively with key stakeholders to help remove shark nets and buoys when particularly large or dangerous swells are forecast, and will immediately alert the shark hotline if a creature is spotted within the nets and/or buoys.

SLSQ also regularly makes use of its Westpac Lifesaver Rescue Helicopter Service to monitor shark movement and assist with shark sightings from the air.

during identified high-risk periods, attack prevalence should substantially decrease. Information on higher and lower risk times and locations has started to emerge (Amin *et al.* 2015; Ferretti *et al.* 2015). The results are showing trends in terms of shark attack risk levels that have the potential to reduce attack prevalence by orders of magnitude – something that cannot be matched by any current strategy that relies on lethal or non-lethal mitigation devices or measures, including culling (Ferretti *et al.*

2015). Furthermore, a greater understanding of shark movement patterns will provide clues into attack motivation. For example, knowing that tiger sharks pup in Hawaiian waters around October suggests that not only will large females be present at that time, but that they may also need to urgently build or replenish energy reserves.

Aerial patrols

Globally, aerial patrols are one of the oldest regional shark attack mitigation methods, and have enjoyed a high level of public and community support. However, the programs have recently come under a fair degree of scrutiny. Aerial patrols are most frequently used for estimating the presence and density of terrestrial animal populations but have also been used for marine mammals, which are generally large and spend a proportion of their time at or near the surface. The detection of aquatic animals from the air is hindered by animal swimming depth and behaviour, water depth, clarity and turbidity, sunlight reflection and the position of the sun, waves, cloud cover and the elevation and bearing angle of the aircraft (Robbins *et al.* 2014; Bryson and Williams 2015).

The purpose of aerial shark patrols is not to conduct formal surveys to quantify the number of sharks or other research activities; instead, they are used to simply detect the presence of potentially dangerous sharks in close proximity to populated beaches for the primary purpose of bather safety. However, data collected from these programs may secondarily contribute to research efforts.

Human observers

Manned aerial shark patrols have been used since 1957 at Illawarra, New South Wales, during peak days and seasons. If a shark is sighted, organisational policies are followed and information on the shark species (often generically broken down into 'hammerhead' or 'other', with 'other' automatically signalling the potential for a dangerous species), shark size and distance from bathers, is immediately relayed to the public via surf lifesavers (on patrolled beaches), the relevant local authorities, and speakers and/or sirens (on unpatrolled beaches). Aerial patrols are capable of surveying large expanses of beaches and are

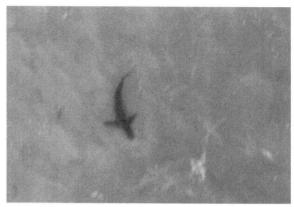

A shark, as observed and photographed from an aerial shark patrol fixed-wing aircraft along the New South Wales coast. Photo by Blake Chapman.

widely recognised as a protective anti-shark safety measure in local communities. However, the difficulty of using aerial surveillance to detect large coastal shark species, which may spend a significant proportion of their time at depth, has been shown to limit the effectiveness of such programs. Indeed, one study using tethered shark analogues to test the efficacy of a trained aerial surveillance crew in a fixed-wing aircraft and an inexperienced surveillance team in a helicopter found that sighting rates were only 12.5% and 17.1%, respectively. The limitations highlighted by that study included animal depth (significant decreases in sightings of animals deeper than 2.6 m) and animal distance from the aircraft (>300 m from the flight path), sun glare, sea condition and observer position within the aircraft (Robbins *et al.* 2014). The short period of individual beach coverage per day (generally only a matter of minutes) was also criticised.

Critiquing authors have presented a quantifiable case on the inefficiency of aerial patrols. But while aerial detection rates seem low, it is arguable that they are no less effective than other mitigation methods that cannot be so easily tested. It is realistically impossible to test the effectiveness of passive mitigation methods (e.g. mesh nets or drumlines), as it is impossible to know how many animals evade those measures. The fixed-wing aerial patrol team involved in the study argued that the detection conditions set by the study were different from normal monitoring conditions (they were set in deeper water and not over white sand, which is generally present along patrolled beaches; Mitchell 2015). Ultimately, though, these statistics, and indeed the even more important statistics that several shark bites in New South Wales waters in recent years occurred in overlapping aerial patrol areas as well as beach meshed areas, support the fact that no mitigation method, or combination of measures, is fail-safe.

South Australia, Victoria and Western Australia also utilise aerial shark patrols. Interestingly, aerial shark patrol flights operate as a tourism activity in Pompano, Florida (http://airtourspompano.com/about/). The latter scheme is particularly clever: as shark patrols would be run anyway, making them a tourist attraction means that the cost of the program can be recouped, even partially, from the paying passengers. Although only a small number of passengers are able to participate in this activity each day (due to the space and weight restrictions of fixed-wing and helicopter aircraft), this type of program could be educational as well, teaching passengers about sharks in the area, shark attack, various mitigation measures and other local ecological topics.

In the future, the use of supporting technology, such as hyperspectral and multispectral scanning equipment, could benefit aerial programs and make them more efficient at detecting sharks.

Unmanned patrols

Unmanned aerial vehicles (UAVs), or drones, are an emerging technology in the field of shark attack mitigation. UAVs encompass all pilotless airborne platforms. These include remotely piloted vehicles (RPVs), unmanned, self-powered, fixed-wing or rotary-wing (helicopter-like) aircraft, and untethered airships (e.g. blimps). There are also unmanned, tethered aerial systems, such as kites and balloons. Unmanned aerial technology has both advantages and disadvantages in comparison to manned aircraft. Some of the advantages include reduced weight, cost (operational and maintenance) and environmental footprint (e.g. emissions and noise), the ability to fly at lower altitudes and slower speeds than manned aircrafts, thus allowing greater flexibility in flight paths, and reduced risk to pilots and passengers. Disadvantages include the current high cost of some aspects of the technology, strict regulations for aircraft operation in the vicinity of populated areas (including public beaches) that therefore require careful planning and operation, the higher accident rate of unmanned aerial systems (UAS) and the related potential risk to people and infrastructure on the ground in the event of equipment failure. A 2005 report showed that UAS accidents were over 100 times more likely than accidents involving manned commercial aircraft (Weibel 2005). This statistic has kept many trials of such systems away from heavily populated areas.

Suitable UASs have been identified for the purpose of shark attack mitigation, but the limitations regarding animal detectability are similar for manned and unmanned technology (Bryson and Williams 2015). Overcoming these difficulties would require more advanced imaging equipment, which would consequently increase the weight of the vehicle (thus decreasing the operation time), and increase the set-up, operational and maintenance costs. Although the use of basic UAS equipment would be cheaper, alterations needed to improve detectability and create systems capable of monitoring entire regions (similar to what is being done with manned technology) would lead to similar operational costs between the two methods (Bryson and Williams 2015). Nonetheless, the technology is already in place to enable unmanned patrol operations, from individual beach monitoring to entire coastlines. As UAS technology is relatively new and the potential for broad-scale application is great, advances in technology and safety, and subsequent cost reductions, will undoubtedly occur rapidly, suggesting that UASs may become a very promising option for shark attack mitigation strategies. Programs are already being developed to automatically recognise and differentiate submerged shapes (i.e. distinguish the shape of a white shark from a dolphin, debris or the aircraft's shadow) from aerial (150 m high) coastal images in uncontrolled natural conditions (Gururatsakul *et al.* 2010). For highly populated beaches, unmanned tethered technology could be a realistic option. Although this technology can cover only a finite area, at particular high-use beaches it could still be advantageous and would be a cost-effective, less regulated and safer option.

Acoustic detection

Researchers across the globe have been using acoustic listening stations to monitor shark movements for decades. Using this method, animals can be tracked over large scales (hundreds of kilometres or more). The low cost and wide scope of equipment deployment in a variety of environmental conditions is a further benefit to this mode of detection.

To be effective for shark attack mitigation, however, several factors must align. First, there must be receivers in the vicinity of the public. With many beaches spread across large expanses of coastline, this could require a large amount of equipment. Due to the popularity of this method in many research fields, there are numerous arrays of fixed receivers already in operation around the world (through organisations such as the Ocean Tracking Network, based at Dalhousie University in Halifax, Canada, which works with collaborators to create a global acoustic telemetry network). Second, the animals can be detected only if they are fitted with a transmitting acoustic tag. Achieving this would require continuous and resource-intensive fishing and tagging efforts, yet these efforts still could not be expected to come close to accounting for all animals that residentially or transiently use an area. Furthermore, deployed transmitters are useful only while they maintain power. Most tags can be set to transmit more or less frequently, to account for various animal speeds and the distance they are likely to travel from receivers. The more rapidly tags transmit signals, the quicker they lose power. Third, receivers can determine the position of a transmitting animal only when that animal is within a certain range (generally limited to ~500 m). Thus, for continuous protection, receivers would have to be deployed at least that frequently over the area that requires coverage. Fourth, the receiving equipment would need to be functioning optimally. This takes into consideration obstacles that may block acoustic transmission (e.g. organic or inorganic objects, or even multiple tagged animals, whose overlapping signals can cancel each other out, resulting in no detections), environmental conditions that limit transmission signal reception (e.g. wave action, salinity, water depth, air bubbles, current, turbidity) or even noise from boat traffic (Heupel *et al.* 2006). Satellite-linked receivers also require regular maintenance. The technology (or human resources) to interpret the signal transmission and relay information to water users must be functional and rapid.

Interestingly, opinion varies as to the desire for this technology. During a media interview, a shark attack victim stated that surfers often prefer not to know if a shark is present before they enter the water (Krajacic 2016). They are conscious of the fact that sharks are in the water and know that they present a risk, but prefer not to always be reminded of their presence. The out-of-sight, out-of-mind mentality lets them enjoy and focus on their activity, instead of continuously worrying about its unlikely risks.

Following the attack on a surfer by an acoustically tagged shark in Western Australia in 2015, the question was asked about whether surface or implanted acoustic tags, which produce bursts of ultrasonic sound, may be hindering a shark's ability to feed (Paddenburg 2016). It has been suggested that acoustic tags may act as an audible warning to receptive potential prey of the predator's approach (Allen 2015). It is still unknown if, how many or which prey species are able to detect the sound of the tags, or if the tagged sharks themselves can detect the sounds, thus presenting an animal welfare dilemma regarding use of this technology. However, products with similar transmitting frequencies are used for the explicit purpose of deterring dolphins. As there is no current knowledge of which species are definitely able to detect such sounds, further investigation is warranted for a complete understanding of how the technology may be affecting the animals and the environment.

Sonar detection

The use of multibeam sonar technology is emerging in the shark attack mitigation field as a close-range, early-warning system. However, the technology and its accompanying compatible, purpose-designed software is still being trialled and modified. At least two products are being marketed in this range: the Clever Buoy (https://cleverbuoy.com.au/#) and the Tritech Gemini SeaTec (http://www.tritech.co.uk/). Trials of the Clever Buoy have taken place at the Abrolhos Islands and Bondi Beach (Australia) but the results are yet to be released, as further trials were deemed necessary.

Sonar systems are used to detect large (>2 m) shark-shaped swimming objects in surrounding water. The current maximum range of detection is ~60 m in deep water, and less in shallow water. The sonar head provides unidirectional real-time imagery with 120° horizontal and 20° vertical beamwidths (Parsons *et al.* 2014). It is unknown exactly how coastal marine conditions (e.g. turbidity, sedimentation, boat traffic) affect detection range and ability. The systems are powered by rechargeable batteries and contain a microprocessor to analyse the data. Sonar data are collected via a transducer anchored to the seafloor and transmitted by an antenna on the buoy's surface. Once triggered, the system sends an 'alert' signal to the shore and over a secure channel to a predetermined audience. At present, sonar systems cannot differentiate between sharks and other large swimming animals, such as marine mammals; however, new sonar image recognition software is being designed to account for shark-specific movement patterns and anatomical characteristics.

Parsons *et al.* (2014) trialled the Tritech Gemini imaging sonar for the purpose of detecting sharks of 1.4–2.7 m. They found that within a 5 m range, sharks were clearly discernible through shape, length and swimming action; however, with greater distance, only movement patterns could be discerned, and then the shark became just an acoustic target. The study identified several limitations of the current technology to detecting the presence of sharks. The most pressing issue was the lack of detection ability in shallow water, which would be most applicable to coastal applications with heavy recreational water user activity. While movement filters can be applied to reduce background noise, any system or environmental movement (e.g. waves, current, surge) would reduce the resolution of actual moving objects. Overall, and even with supporting modifications, the technology was deemed to be imperfect and left relevant gaps in shark detection potential.

Sensory barriers

A wide variety of products have been marketed and trialled that aim to deter sharks based on their sensory modalities. In many cases, the background concepts of regional barriers are similar to those used in personal devices – regional mitigation methods are simply developed on different scales to facilitate protection over a much greater area.

Electrosensory deterrents

Although not commonly used for regional (beach) mitigation measures, research into electrical shark deterrents has been going on for decades.

The first trials were initiated in South Africa in the late 1950s and the first electrical cable was deployed in 1972 (Cliff and Dudley 2011). The studies were eventually abandoned due to difficulties in maintaining the equipment, in operating in rough environmental conditions and in determining the effectiveness of the cables (Dudley and Cliff 2010). More recent successes in personal electrical shark deterrents are being used in studies of larger-scale technology for beach-wide use, and several private companies have begun exploring the creation of electrical or electromagnetic beach barriers.

The main challenges for beach-scale electrical devices are the amount of electrical potential that would need to be generated to cover such a large area, and how to contain the electrical potential (and associated equipment) in such harsh and dynamic environments to the degree that it would be effective. Another major challenge is that, while many species show sensitivities to electrical fields, behavioural responses among species can be quite variable – what may deter one species could attract another (Haine *et al.* 2001; Kajiura and Holland 2002; Jordan *et al.* 2011). It is expected that humans in close proximity to such devices would feel some effect from the electricity, although this would likely be felt only as a tingly sensation. More problematic would be the risk of potential interaction with life-sustaining electrical implants in recreational beach users, such as pacemakers – something that needs to be investigated further in relation to specific electrical devices.

The Sharksafe Barrier is currently the only regional shark mitigation technology that employs magnets, specifically grade C9 barium-ferrite permanent magnets. However, the more dominant aspect of this product is the concurrently employed visual barrier made of PVC pipes and designed to resemble kelp. The combination of sensory deterrents means that the technology may work even in conditions where one sense is compromised, such as the visual system in turbid or murky water. Open-water trials of the Sharksafe Barrier have shown that shark behaviour can be manipulated by the barrier and that white sharks do not swim through the barrier, even when food is placed within its confines. However, like other barriers, the current Sharksafe Barrier model is only 15 × 15 m. There is nothing to prevent sharks from simply swimming around the barrier, and indeed they did so during trials, passing only metres from the barrier (O'Connell *et al.* 2014). Biofouling could also be problematic for this technology.

Visual deterrents

Visual deterrents aim to discourage a shark from entering an area, while not physically preventing it from doing so. Certain visual barriers have the advantage of not requiring a supporting power supply or service checks for animal entanglements. In practice, visual barriers are often combined with other technology to boost effectiveness, as with the Sharksafe Barrier. In that equipment, the PVC pipes are anchored to the bottom but are free to move in the current, creating a fluid effect that is theorised to help prevent sharks' habituation to the barrier.

Other visual deterrents under investigation include strobe lights and bubble curtains, but neither has been comprehensively tested at this stage. The idea behind strobe lights is that the flicker of the light would be set to a rate of maximal contrast sensitivity (although this would vary somewhat between species), thus confusing or irritating the sharks' visual system. Bubble curtains work simply by pushing air through a submerged perforated hose to produce a wall of visually obvious bubbles. This type of barrier has been trialled and used in aquariums to prevent sharks from running into walls, or for sectioning off particular areas of a tank, with mixed (generally species-specific) results. In addition to the clear visual component, the air being pushed through the perforations in the hose produces hydrodynamic cues that could be perceived by the shark's auditory or lateral line systems. The benefits of this sort of technology are its simplicity of design, potential for use in a range of locations, the lack of mortality in target and non-target species and the relative ease of maintenance. However, its effectiveness as a regional shark deterrent is yet to be determined.

Case studies of regional mitigation practices
Western Australia

Western Australia has had one of the most active and fluid shark attack mitigation campaigns in the world for some years (DPC 2014, 2016;

McAuley *et al.* 2016). The campaign has involved several different programs and has been revised, reworked and had supplemental measures employed every couple of years. The measures were initially deemed necessary in response to 10 deaths from shark attacks in state waters over a 10-year period (in stark contrast to the 20 deaths recorded over the past 100 years). More recent measures have been taken in response to the even more concerning seven deaths in the three and a half year period up to 23 November 2013. Overall, AU$33 million has been spent or committed to the program since 2008, and a broad range of mitigation measures and actions have been investigated, trialled and reviewed. The overall and most specific aim of the program is to catch potentially dangerous sharks that come close to popular beaches and surfing locations during the summer months (October–April).

The mitigation measures have included:

- December 2007–July 2015: acoustic tagging of 50 white sharks, 55 bronze whaler sharks and 70 tiger sharks;
- December 2008: funding for helicopter patrols by Surf Life Saving WA (AU$2 million/year);
- 2009–2012: Shark Monitoring Network project (metropolitan-only feasibility trial between 2009 and 2011 and an operational data-provision system in 2012);
- November 2011: development of a shark response unit within the Department of Fisheries;
- 2012–2020: implementation of the Serious Threat policy (formerly the Imminent Threat policy) (AU$2.8 million);
- 2012–2016: research funding (AU$3.6 million) to investigate:
  - ➤ possible links between shark interactions and environmental conditions;
  - ➤ white shark population estimates;
  - ➤ the effectiveness of netting as a shark hazard mitigation measure;
  - ➤ non-lethal shark detection and deterrent technologies;
- July 2012: discussions on regulations to rule out tourism ventures based on attracting sharks (e.g. cage-diving operations);

- December 2012: funding to research shark hazard mitigation strategies;
- December 2012: jet ski (12) deployment by Surf Life Saving WA for beach patrols (AU$1.2 million);
- 2013–2020: Shark Monitoring Network and shark tagging and tracking (AU$3.7 million);
- January 2013: new watchtower at Cottesloe Beach (AU$300 000);
- August 2013: launch of The BeachSafe smartphone app;
- October 2013: beach enclosure trial at Dunsborough;
- December 2013: funding for four shark mitigation research projects (AU$967 000);
- January–30 April 2014: drumline trial in two Marine Monitored Areas (AU$1.28 million);
- January 2014: launch of the Sharksmart website;
- 2015–2016: four additional beach enclosures (AU$1 million).

2007–2015 acoustic monitoring program

Acoustic receivers have been used in Western Australia to monitor shark occurrence, movement and behaviour since 2009. This system, known as the Shark Monitoring Network, utilises up to 334 acoustic receivers that span from the south-west part of the state northward to Ningaloo Reef (McAuley *et al.* 2016). Between 2007 and 2015, 50 white sharks were acoustically tagged in Western Australia, plus an additional 151 in South Australian waters. A further 55 bronze whalers (*Carcharhinus brachyurus*) and 70 tiger sharks were also tagged. The tags transmit every 50–150 seconds, and can be picked up by receivers within a 400–500 m radius. The Shark Monitoring Network also utilises 25 satellite-linked receivers that provide near-real-time monitoring of the sharks' acoustic activity around some of the most populated beaches. This information is sent to public safety officials and to the public through social media networks and dedicated websites. Many of the sharks tagged early in the program were fitted with external tags, which reliably transmitted for a maximum of three years. Later efforts focused on the use of internal tags, which are capable of transmitting for close to a decade.

Interestingly, data from the Western Australian monitoring network revealed that of the 201 white sharks tagged in Western Australia and South Australia, just 64 have been detected through acoustic monitoring. The sharks were over 10 times more likely to be picked up by receivers off-shore, than by metropolitan beachside receivers. Tagged sharks commonly transmitted in deep water (94% were at depths of more than 50 m) and far off-shore (88% more than 10 km off the mainland coast), suggesting that most white sharks in the region would not present a threat to most water users (McAuley *et al.* 2016). Further results showed that only three white sharks were detected in the region more than a year after they were tagged, again proving that, for acoustic tagging to remain an effective mitigation measure, continuous tagging efforts would be needed to monitor current regional populations of such mobile animals. Bronze whalers, which can also pose a significant threat to human water users, were found to be more residential.

2012–2014 Imminent Threat policy

In November 2012, the Western Australian Government issued an Imminent Threat policy, which enabled the killing of individual sharks, regardless of species or conservation/protection status, within state waters (which generally extend 3 km from shore) if they were deemed to pose an imminent threat to public safety. The most basic defining characteristics of 'imminent threat' were where a fatal shark attack had already occurred and a relevant individual shark remained in the area. However, the policy also stipulated that a shark could be killed without the occurrence of a fatal attack during a situation of high hazard and high risk. This was defined as following a sighting of a potentially dangerous shark (high hazard) in close proximity (generally within 1 km) to popular beaches, during daylight hours and in conditions conducive to overlapping human beach use, and when measures to clear people from the water, and keep the area clear from human use, were deemed to likely be ineffective (high risk) (DoF 2012a). Only authorised contracted fishers or appropriately trained personnel with suitable equipment were permitted to carry out the capture and destruction of animals, when it was deemed safe to do so, and only in the event that they could respond to the location of the sighting within one hour. Other measures had to be employed simultaneously, such as beach closures within the proximity of the deployed capture gear.

In late 2014, the policy was revised to become the Serious Threat policy. The amended policy guidelines (which included little change) were published in January 2015 and are still in effect. Since its inception, the policy has been implemented eight times. One followed a non-fatal attack in Esperance in 2014, where two white sharks were caught and destroyed, and another resulted in the destruction of a single shark near Falcon in 2016, following the death of a surfer. In both instances, the government was unable to find evidence that it had caught the offending shark. The other six enactments failed to catch sharks. Even though an order is made at the discretion of the government, the hunt-to-kill policy is said to be carried out when 'the public are less concerned with being measured, wanting instead reassurance, even revenge' (Young 2017).

2014 drumline program

Under this program, up to 60 static baited drumlines were deployed and set ~1 km off-shore from popular metropolitan and south-west swimming and surfing beaches. The lines were monitored for 12 h/ day, seven days a week. The equipment was designed to target white sharks, tiger sharks and bull sharks over 3 m in total length, which were deemed to be the individuals of greatest concern to public safety in the region.

A total of 172 sharks were caught. Of these, 50 were ≥3 m tiger sharks. No white sharks were caught. The equipment used in the program was designed to minimise bycatch and environmental impact through the use of 25/0 circle hooks, the removal of lines before the commencement of seasonal coastal whale migrations, the deployment of hooks at a depth great enough to reduce interaction with sea birds, and daily monitoring of the lines. These measures were considered effective, as only eight non-sharks were caught (5%), including seven rays and one blowfish. Sharks <3 m and still in suitable condition at the time of collection were tagged, photographed, sexed, measured and released, and the information was publicly available through the Surf Life Saving Western Australia Twitter feed. Target species >3 m in total length were killed with a firearm. All dead animals were transported off-shore and disposed of.

During the program, three sharks that had been acoustically tagged were detected by receivers in close proximity to baited drumlines; however, they were not caught, proving that the drumlines were ineffective in catching all the target animals in the area.

Despite the considered development of the program, it was heavily criticised by many. The most significant criticism was that the program was not based on science. Other significant criticism focused on the extent of misinformation received by the public and the disproportionate media coverage. Ultimately, in terms of shark attack mitigation, the trial was deemed to have been too short to have generated sufficient data for measuring a reduction in risk. A submission requesting an additional three-year implementation of the program was retracted, and similar measures have not been employed since.

Beach closures

Currently, Surf Life Saving Western Australia maintains the following beach closure regime (http://surflifesavingwa.com.au/safety-rescue-services/shark-safety):

In the event of a shark sighting, the following precautions are implemented:

- **If the shark is larger than 3 m and within 1 km of the shore OR if the shark is 2–3 m in length and/or schooling sharks and within 500 m of the shore**: Close the beach and water 1 km either side of the shark location for one hour (two hours if at dawn, or for the remainder of the evening if at dusk);
- **If the shark is less than 2 m in length**: Advise the public but maintain normal operations.

Recife, Brazil

Following an abnormally high regional shark attack rate beginning in 1992, a Decree (No. 18.313) was published in Pernambuco state

(Brazil) in 1995 establishing a ban on surfing, body-boarding and other marine-based activities in areas within the maritime border of the state. However, shark attacks continued, and in 2004 the state government implemented Decree No. 26.729, which formed the Committee for the Monitoring of Shark Attack Incidents (CEMIT) under Article 37, Items II and IV of the state constitution, to operate within the scope of the Department of Social Defence. The committee was implemented, considering the constitutional duty of the state, together with the municipalities, to adopt actions aimed at:

- protecting and defending the health, physical integrity and well-being of the population;
- the known presence of sharks in certain areas of the state of Pernambuco and the need to develop awareness, guidance and educational work in partnership with society, regarding knowledge, precautions and coexistence with sharks;
- the need to adopt measures to monitor and investigate shark incidents in the risk areas off the coast of Pernambuco, aimed at the collection of scientific data to enable decision-making to reduce the probability of those occurrences;
- the need to seek integrated solutions that allow for the immediate reduction of shark attacks, in harmony with necessary environmental balance;
- the measures proposed at a symposium held on the subject.

The purpose of the committee was to propose measures to monitor and investigate the presence of sharks in particular areas of the Pernambuco coast, in order to minimise shark attack frequency and provide adequate information, guidance and public education. The committee was initially composed of formal members from the Secretary of Social Defence, Military Fire Brigade of Pernambuco, Institute of Legal Medicine, Federal Rural University of Pernambuco (who served as President, by delegation), and the Department of Fisheries Engineering of the Federal Rural University of Pernambuco – Instituto Oceanário de Pernambuco. Representatives from a variety of other legislative assemblies, recreational organisations and associations, tourism industries, research and teaching institutions and the press were invited to participate as guests. The specific duties of the committee were outlined as:

- monitoring and recording incidents with sharks, with statistical data;
- defining strategies and actions that aim to minimise the risk of attacks on affected beaches;
- following the actions taken by the various agencies, related to incidents with sharks;
- acting as a reference centre, to guide information and discussions;
- evaluating impacts of all kinds, whether economic, social or environmental, arising from incidents and actions.

The Decree was modified (No. 41.251) in November 2014 to amend the committee's membership to the Secretariat of Social Defence, through the Military Fire Brigade of Pernambuco, Military Police of Pernambuco and Institute of Legal Medicine, and the State Agency of the Environment.

Overall, the committee was designed to address various components of the shark attack problem through a multidisciplinary approach. One of the initial scientific research goals was to design a non-lethal solution to the problem, and the Shark Monitoring Program of Recife (SMPR) was developed to relocate potentially dangerous sharks from areas deemed to be overlapping with human use (not remove those sharks from the population) (Hazin and Afonso 2014).

In the SMPR, sharks deemed to be potentially dangerous are caught, transported and released off-shore (Hazin and Afonso 2014). The program utilises a protective fishing method, with a combination of bottom longlines and drumlines. Both longlines and drumlines utilise circle (17/0, 10% off-set) hooks. Bait typically consists of moray eels (*Gymnothorax* spp.) (~300 g) and occasionally oilfish (*Ruvettus pretiosus*). Two longlines (100 hooks each) are deployed alongshore ~1.5–3 km away from the coastline at each site, at average depths of 13.2–13.5 m. A total of 10–13 drumlines are deployed ~0.5–1 km from shore, at depths of around 6.4–10.2 m. The longline configuration is set to form an outer barrier for intercepting sharks before they enter areas of risk and to increase the probability of sharks being lured away from the channel by the bait. The drumlines provide a secondary barrier for sharks entering the channel. Longline soak time is generally 14–15 hrs and drumlines operate continuously with bait replacement each morning.

Potentially dangerous species are transported in a ~3 × 1 m tank on the deck of a 13 m research boat. During handling and transport, measures such as altering the gunwale of the boat for easy removal so the shark could be hauled directly from the water into the tank, the placement of soft, impact-absorbing material underneath the shark to minimise internal damage and covering the animal's eyes with dark tissue are taken to minimise the risk of injury to the animal. All other species are released at the site of capture. Fishing operations are conducted continuously from Fridays to Tuesdays to overlap fishing effort with weekly peaks in recreational beach usage.

Over an eight-year study, the SMPR showed a high selectivity for sharks (compared with shark meshing) but also included other fish and turtle bycatch. The program targeted Carcharhinid and Sphyrnid (hammerhead) sharks; however, these species only accounted for 7% of the total catch. The fishing mortality of most species was generally low, except for blacknose sharks (*Carcharhinus acronotus*) and moray eels. Protected species had ~100% survival.

Scientific review of the program deemed it to have been very successful, citing it as a major factor in the 97% reduction in the regional shark attack rate during the program's active operation, compared to program inactivity. Surfing was banned in Recife during the study period, which may have also contributed to the diminished shark attack rate.

Réunion Island

After five shark attack fatalities between 2011 and 2013, the authorities of Réunion Island initiated a program to kill 90 tiger and bull sharks. Bans/closures on certain beaches and water activities were also implemented. In 2012, a pilot study was conducted by the Surf League to trial the use of surveillance and intervention teams on and under the water when organised activities were held. Further experimental protocols and analysis have been conducted in an effort to evaluate the effectiveness of such a program. Two of the island's beaches, Boucan Canot and Roches Noires, are equipped with shark nets.

Following what was deemed to be a crisis situation in April 2015, eight initiatives were announced to accelerate and extend the implementation of the shark risk reduction plan at Réunion Island and the budget for this strategy was raised to €6 million for the period of 2015–2020. The eight initiatives included:

- increased effort to secure bathing and water activities, including strategies such as 'reinforced shark watches', safety nets and other techniques that have demonstrated effectiveness;
- a controlled rise towards a 50% increase in targeted tiger and bull shark fisheries;
- the search for sustainable solutions to commercialise shark catches (e.g. research into the quality of the meat for human consumption, currently banned due to the risk of ciguatera contamination);
- the preservation of the marine ecosystem within the Marine Nature Reserve in conjunction with the controlled shark fishing programs;
- intensification of scientific research projects, in particular relating to extending the network of listening stations, increasing shark movement monitoring programs and evaluating shark population numbers on the outskirts of Réunion Island;
- strengthened support for prevention groups whose initiatives are involved in the global response of the prevention and reduction of risks (e.g. surveillance and prevention programs and water quality monitoring);
- accelerating the establishment of the Resource and Support Centre intended to coordinate the shark risk reduction policy, communicate to the public and perform crisis management;
- the restoration of tourism in Réunion Island.

Public perception of shark mitigation measures

Beach users have been found to be moderately aware of mitigation measures employed in their regions; however, the perceived efficacy of those programs is often overestimated. More than half of a surveyed

South Australian population believed that aerial spotters spent 30 minutes over each beach per day and spotted 50–75% of sharks. The actual rate was just 17%, with individual beach observation times of just a couple of minutes. Also, ~40% of respondents believed that shark nets were in operation, although they are not in South Australia (Crossley *et al.* 2014). It has further been found that beach users generally do not select beaches for recreational use based on the shark attack mitigation measures employed there.

Overall, feelings on sharks and shark attack mitigation measures are extremely mixed. Many people still demand the destruction of sharks implicated in attacks, non-selective culls or even the destruction of entire (or multiple) species of sharks. However, even with increasing global shark attack incidence, this attitude seems to be less prevalent. A 2013 survey of 557 Western Australian ocean users found that 34% of respondents strongly opposed the Western Australian Government's shark hazard mitigation policy (implemented in 2012), and a further 19% opposed it (Gibbs and Warren 2015). Yet, a noteworthy minority (30%) supported it. Only 17% of respondents felt that the policy would reduce the risk of shark attack injury or fatality, and just 8% agreed that the policy provided them with more protection and confidence in the water. Respondents felt that the policy would be very unlikely (16%), unlikely (20%) or neither likely nor unlikely (39%) to bring tourists to the region. Two independent surveys (including one commissioned by the Western Australian Government) corroborated those findings. A survey conducted by UMR Research was reported by the media to have found that the majority of people (83%) have not changed their ocean usage as a result of the threat of shark attack and 82% of people were against killing sharks, in the belief that people enter the water at their own risk (Dorling 2014). The 2013 Marketforce survey (Marketforce 2013) commissioned by the government also found that the majority of people had not changed their beach usage over the previous two years; among those who had, less than 6% did so because of the threat of shark attack. Despite mitigation efforts, 40% of people (in 2013) felt that they were more likely to encounter a shark than they were two years previously, suggesting a lack in confidence of government-implemented mitigation measures. Most individuals surveyed (53%) felt more needed to be done in relation to the risk of shark attack, but that culling was not the way, 29% felt nothing needed to be done and 19% felt culling was the best approach. The majority (59%) of respondents felt that beach safety, in terms of shark attack risk, was up to the individual; 26% felt that it was up to the Western Australian Government. Overall, public perception survey results perfectly highlight the large divide in public opinion and the extreme complexity of shark attack mitigation management.

Many people felt that the Western Australian Government's Imminent Threat policy was a kneejerk reaction, as it was not based on or supported by scientific evidence or proof of concept in regards to reducing human injury or fatality. Strong, overt public dissatisfaction was also shown towards the 2013–2014 drumline program, as the following critical responses demonstrated (DPC 2014):

- a threat of legal action by a tourism operator (which was not pursued);
- legal action by an environmental lobby group and family member of a shark attack victim (which was decided in favour of the government);
- an appeal against the above decision (subsequently withdrawn);
- a third-party referral to the Environmental Protection Authority, which resulted in over 20 000 submissions and a decision not to subject the program to an environmental assessment;
- a petition tabled in the Western Australian Parliament;
- two protest rallies at Cottesloe Beach, attracting thousands of participants;
- 12 Freedom of Information requests;
- 28 Western Australian Parliamentary Questions, many with multiple subquestions;
- 765 articles on sharks in local, state and national newspapers;
- 1100 radio news bulletins on sharks (Western Australia);
- 850 radio talkback comments on sharks (Western Australia);
- 290 television news items on sharks (Western Australia);

- 286 000 emails and letters to the Department of the Premier and Cabinet on sharks (a significant number of which were pro forma emails);
- a significant number of postings on Twitter and Facebook about the drumline strategy (some of which were offensive and contained personal attacks on members of the government and staff involved with the program).

Indeed, the two Cottesloe protests attracted 10 000 people between them, plus an additional 2000 at Manly in Sydney and others at 11 further locations around Australia (Gibbs and Warren 2014). The use of social media (Twitter and Facebook) and online petition sites led to an influx of submissions to the Environmental Protection Agency, opposing the cull.

Surveyed recreational water users were found to be generally against lethal (or potentially lethal) management strategies, particularly baited drumlines, culling of potentially dangerous species, increased usage of shark nets, and track, catch and kill methods (Gibbs and Warren 2014). The most supported strategies were an improvement in public education about sharks, encouraging acceptance of sharks in natural waterways, and increased warning systems. Ironically, these opinions largely contradict many of the mitigation strategies employed by governments around the world. The call for culling to be stopped in consideration for the wishes of the person bitten is currently the focus of FIN FOR A FIN™ (https://www.finforafin.com). Through this initiative, surfers can attach a specially marked fin to their board as a physical and visual alert, signifying that if they are seriously injured or killed by a shark, they do not wish retaliatory action to be taken against the animal. They can also register their name on a global database, ensuring their wishes will be shared with relevant authorities in the event of an attack.

Conclusion

Analysis of current shark attack mitigation technology shows that no measures have proven to be effective if employed alone. Considering the great variety in shark biology, physiology, ecology and regional

FIN FOR A FIN™ was started to break the cycle of vengeful killing that often follows a shark attack.

Following extensive discussions with surfers and shark attack survivors, we know that a love and respect for the ocean is common among these people. They don't want a shark killed in the event of an attack, but can't always make their wishes known if the attack proves to be extremely serious or debilitating.

As a group of creative thinkers – some of whom surf but all of whom love the ocean and admire and respect sharks – we wanted to get involved in the conversation and try to find an alternative means of protecting surfers and sharks. Noticing the parallel between the shark fin and the surf fin, the idea was to design a high-performance fin that any surfer would appreciate and use, that would also act as a visual symbol to promote the coexistence of surfers and sharks and alert authorities to not kill the shark in the unfortunate event of an attack, as often the surf fin may be all that survives.

Although started in Australia, we believe that this can be much bigger. We hope that governments will recognise this initiative and respect the surfers' wishes. We know that policy won't change overnight, but we hope to work with governments in the future to find the best way(s) to protect the lives of surfers and sharks.

FIN FOR A FIN Team

oceanography and human water usage, it is obvious that a single solution will never be effective in fully mitigating the risk of shark attack. Since the first shark attack mitigation devices were conceptualised more than 60 years ago, our understanding of sharks, as a whole, has grown remarkably. We have a much greater understanding of shark sensory capabilities, movement patterns and physiology, all of which are necessary for the development of effective mitigation measures. Public demands have led to a focus on technological advances and mitigation practices that limit both target and non-target species mortality, and very few long-term mitigation policies/measures have remained stagnant over the years. Both existing and new shark control programs have begun to favour more advanced, environmentally friendly and scientifically backed approaches.

Although resource-intensive, the effectiveness of human monitoring programs cannot be understated. This has been proven in general water safety through the use of lifeguards, and more specifically in relation to shark attacks with the Shark Spotters program in South Africa.

Although the Shark Spotters program relies on particular geographical and environmental attributes for functionality, the concept of increased monitoring and awareness is solid. An advanced monitoring and rapid response framework of some sort would likely be very beneficial for locations like Réunion Island that have relatively concentrated affected areas, but high shark attack fatality rates. Tethered aerial surveillance technology connected to programs designed to autonomously distinguish shark shape and movement patterns for high-use beaches also seems to have potential for real-time shark monitoring.

New or continued use of traditional lethal mitigation measures is not recommended, but the evolution of some of these programs may lead to promising future measures. For example, the use of appropriately resourced SMART drumlines could be promising. At the very least, simple modifications to existing set-ups, such as switching from J hooks to circle hooks and the implementation of more frequent checks of lines, would be beneficial. This has been demonstrated in the drumline capture and relocation program in Recife, which has produced compelling results with limited (immediate) animal mortality. While the long-term survival rate of hooked animals has not been reported, it could be easily measured through tagging trials. Such trials would not only quantify long-term survival, but also confirm the movement patterns of potentially dangerous species to verify whether they return to populated beaches following release.

In many cases, the most important outcome of shark attack mitigation measures is not to remove sharks, or even alter their behaviour; instead, it is more about regulating, calming and placating the general public (Neff 2012). Thus, some of the most necessary mitigation measures may, in fact, need to focus on human behaviour and management, as opposed to shark control. This is supported by public opinion, which has suggested the need for greater public education to understand sharks and shark attack. It is also important that the public is educated well enough (with accurate information) to be more heavily engaged in policy-making.

Finally, continued research into understanding the behaviour, distribution, population status, movement and ecological role of sharks, in addition to motivational factors that may increase the likelihood of shark attacks, is imperative in reducing the risk of negative shark–human interaction. However, we need to recognise the limitations of scientific research. Conducting, analysing, documenting, publishing and communicating research often takes years from start to finish, so findings may paint a retrospective picture of a situation. Some research only mimics natural conditions or encounter situations, thus not incorporating the many complexities (e.g. variable environmental conditions and shark motivation) of actual attacks. I certainly don't want to downplay the importance of research but, as with every other aspect of shark attack mitigation, there are constraints that need to be understood and considered. Further research into shark biology in general is also needed, specifically in sensory biology (as this is often exploited for mitigation purposes) and in physiology and reproductive patterns.

10

Legislation relating to shark attack mitigation

Statistically speaking, the probability of shark attack falls into an unusual category of risk management, known as the zero-infinity problem. This term relates to situations where there is an almost zero probability of an event (e.g. shark attack) occurring, but nearly infinite consequences to the victim, those close to them, the community and the local tourism economy when the event does occur (Caldicott *et al.* 2001). Yet, to the human psyche, statistics on the matter are irrelevant to levels of perceived concern. The likelihood of the loss of human life or severe injury as a result of shark attack is extremely low but humans are not rational beings and, especially when it comes to matters of survival, we overemphasise events that could compromise our health and safety. The media compounds such issues through an onslaught of coverage and its use of often-sensationalised language. Surveys of the public have shown that policy and decision-making on shark attack mitigation action (or lack thereof) can be politically debilitating. The range of opinions held by individual members of the general public are so widely varied, based on their previous experience relating to sharks (and the marine environment in general) and prior beliefs and education, that no matter what decisions are made a significant percentage of the population will be unhappy and (often vociferously) opposed to those decisions. Legislation relating to shark attack and indeed, more broadly, human–wildlife conflict management on the whole, has become as much of a political challenge as it is a scientific one. Voters commonly punish governments for 'acts of God' such as natural disasters, droughts, floods and other events that are out of their control, such as terrorism and shark attacks. While governments cannot be held accountable for such events, they can be held accountable for the actions they choose to take in response. The main challenge that governments face in dealing with these sorts of issues is the formulation of policy based on fact, not on perception or emotion.

What constitutes legislation

In many democracies, legislation is based on a hierarchical system of legally binding guides that inform the public how to operate in a particular field. These guides take the form of Acts (laws), amendments, regulations or codes, and are legally enforceable. While enforceable legislation relating to sharks exists in many places around the globe, it almost entirely relates to sustainable practices and conservation, and mostly revolves around restrictions and regulations for commercial and recreational fishing and finning practices. Legislation specifically relating to shark attack mitigation is rare. The development and implementation of policies, strategies and guidelines is a far more common governmental response. These non-legislative instruments function as administrative addendums to support regulation (and must be authorised by, or at least consistent with, relevant legislation), but lack independent legal authority and can be developed and approved by various officials and at different levels of governments. While this allows for more bespoke, regionally specific responses, it also results in uncoordinated, inconsistent and often misconceived shark mitigation measures. As these measures do not comprise formal legislation, policies, strategies and guidelines relating to shark attack have been termed 'mitigation measures' in this book, not 'legislation'.

Legislation relating to shark attack

Shark attacks are ungovernable, and there are no simple, effective and symbiotic solutions that governments can implement in response to sharks biting people. Shark attacks occur in the natural world. While governments can aim to reduce the incidence of shark attack through various mitigation measures, educational initiatives or bans on recreational water use activities, the decision to share an environment with sharks is made solely by humans. Regardless of the risk, and

214

occasionally regardless of related policies or legislation, there will always be people who insist on entering natural waterways.

In what is assumed to be an attempt at humour in an article relating to the politics around shark attack, Achen and Bartels (2004) included a footnote that said:

> On 17 December 1967 Australian prime minister Harold Holt disappeared while swimming in shark-infested waters at Cheviot Beach near Portsea, Victoria. His body was never found. Being devotees of democracy, however, we disapprove of this apparent attempt by the sharks to cut out the middleman.

Although perversely phrased (and not a confirmed shark-related fatality), the point remains that shark attacks are a 'punishable offence' for politicians. As a result, governments attempt to make shark attacks governable; however, they often respond to the emotional response of the public and the media, instead of the reality of the event. In doing so, they shift the event from the natural world into the social world. Governments engage in the topic and they introduce policies. While opinion is divided on whether that is the right course of action, when governments become involved in the rhetoric that assumes the element of intent, bites are termed 'attacks' and natural animal behaviour is criminalised, the magnitude of the situation becomes out of proportion and inappropriate. 'These are terrible events. But governments have a choice [in how they discuss shark–human interaction]. They have educated the public that sharks are movie monsters that need to be killed' (Christopher Neff in Young 2017).

In order to learn more about the politics that surround shark attack, and how governments are addressing the matter, I contacted government representatives from the 10 countries/territories that have the highest shark attack prevalence – the US, Australia, South Africa, Brazil, New Zealand, Papua New Guinea, Réunion Island, Mexico, the Bahamas and Iran. I also contacted certain regional governments that are most specifically affected by the occurrence of shark attack. These included the state (or equivalent) governments of Florida (US), Hawaii (US),

State governments appropriately have primary responsibility for public safety and managing risks from sharks. The Minister for the Environment and Energy (the Minister) has a legislated role to ensure that shark hazard mitigation activities are consistent with national environment law, the *Environment Protection and Biodiversity Conservation Act 1999* (EPBC Act). The Minister is committed to working with states to find a path that achieves both public safety and good outcomes for our environment.

The EPBC Act protects matters of national environmental significance. Matters of national environmental significance include nationally listed threatened species and ecological communities and listed migratory species, among others. A person proposing to take an action that is likely to have a significant impact on a matter of national environmental significance must refer their proposal to the Department for assessment and approval. Substantial penalties apply to a person who takes such action without approval.

Under the EPBC Act, the Minister may exempt certain provisions of the Act from applying to proposed actions if he considers it to be in the national best interest. The Minister recently made such a decision in November 2016 in response to an application concerning the NSW North Coast Shark Meshing Trial. The exemption is for a limited time and was granted to enable the NSW Government to commence the trial immediately in anticipation of the peak holiday and tourism season when the beaches are most heavily used. A statement of reasons for this decision is publicly available on the Department of the Environment and Energy website at: http://epbcnotices.environment.gov.au/exemptionnotices/.

The Australian Government is also working to better understand sharks and has recently provided funding (about AU$1.5 million over four years) for additional research into white sharks. This research will improve the management of this important species. You can find more information on shark species and how they are managed under national environment law via the Department's website at: http://www.environment.gov.au/marine/marine-species/sharks.

Kim Farrant
Assistant Secretary
Department of the Environment and Energy
Assessments (NSW, ACT) and Fuel Branch

New South Wales (Australia), Western Australia (Australia), Queensland (Australia) and Pernambuco (Brazil). I requested a response, on behalf of the government, detailing the respondent's personal and political view of sharks and shark attack in their region, and a summary of any previous, current or proposed legislation relating to shark attack. The only response on behalf of a country came from the Australian federal government. However, responses were also sent on behalf of the Hawaiian, New South Wales, Western Australian and Queensland state governments.

Historical precedent for shark attack legislation

In 1916, 'the most serious string of shark-related fatalities in American history' occurred in New Jersey, when four people were killed and another injured over a 13-day period within a span of less than 150 km (Achen and Bartels 2004). These were the first shark attacks ever recorded for the state. The first attack was largely ignored due to scepticism about whether unprovoked shark attacks occurred at all, let alone in waters as far north as New Jersey (Fernicola 2001). However, as the fatality count grew, word spread and the affected towns lost significant tourism visitation and income during what should have been their peak season. Consequently, the series of events led to the first call for governmental action in relation to the threat of negative shark–human interaction. A cabinet meeting was held to discuss the matter, but no effective measures were devised or enacted (Fernicola 2001). Shortly after the final attack, the recreational beach and tourism season ended anyway, so no further effort was expended. Retrospective analysis of voter preference, however, suggests that the events did affect voting in the presidential election that occurred less than four months later, but only in the seaside towns that had lost tourism revenue as a result of the attacks (Achen and Bartels 2004).

The first legislated action relating to shark attacks occurred in New South Wales, Australia, in November 1929, in response to 13 shark-related incidents and seven fatalities in the state. A £10 fine could be imposed on 'disorderly' swimmers, termed 'shark bait' because of their actions of swimming at dawn, dusk, alone or far from shore. Notably, imposing fines on beach users attests that human activity was considered at blame for human–shark interactions at the time (Neff 2012).

Current legislation relating to shark attack

More recent legislation still incorporates a degree of human accountability in negative shark–human interaction, although blame is generally attributed to the unpredictability of sharks. On 6 January

1995, Decree No. 18.313 was published in Pernambuco (Brazil) establishing a ban on surfing, body-boarding and other marine-based activities in areas within the maritime border of the state. The decree was implemented on the basis of an abnormally high rate of sharks 'victimising the surfers in certain areas of the sea coast of the state'; an urgent need to institute appropriate disciplinary action and enforce the policing of surfing (and similar activities) in areas of 'imminent risk'; the constitutional responsibility of the state and other federated entities to adopt measures aimed at protecting and defending the health, physical integrity and welfare of the population; and the main and urgent goal of reducing, as far as possible, the 'alarming statistic' of shark attacks, especially in relation to surfers and those engaging in recreational activities in the state (Estado de Pernambuco 1999). The execution of the decree falls to the Military Fire Brigade, which is responsible for providing information to the public on the dangers of the prohibited areas and the subsequent restrictions; supervising the prohibited areas and enforcing the prohibition of nominated activities; seizing the boards and equipment of individuals who violate the prohibition; and extending the concepts to similar activities and events. The state provides signage, guidance and clarification to the public about the potential risk of surfing, body-boarding and other marine-based recreational activities. The initial legislation has been modified to amend and clarify the jurisdictional area and to more clearly define the activities prohibited, but overall, the ban still stands.

Swimming and surfing are generally banned within a 300 m perimeter zone off the Réunion Island coast due to the threat of shark attack. Non-motorised activities (e.g. paddleboarding and kayaking) and snorkelling, bathing and swimming are authorised without specific restriction only in certain lagoons, including l'Ermitage, La Saline-les-Bains, Trou d'Eau and Saint-Leu (west coast) and Saint-Pierre (south coast) (A. Payet, *pers. comm.*).

As with non-legislated mitigation measures, the quantifiable effectiveness of restrictions and bans on water-based activities is extremely difficult to ascertain. Annual unprovoked attack rates have remained low in Brazil following the instigation of numerous mitigation measures, including bans on water use activities. However, the effectiveness of the entire suite of regional measures can only be evaluated together; it is impossible to separate and assess the effectiveness of a single component. Although Réunion Island banned water-based activities along much of its coastline in 2013, the attack rate has remained elevated and the island reported the second-highest number of attacks in a single year (four) in 2015, following the six attacks in 2011. While attack prevalence may well have been even greater without legislative bans, it is impossible to say so with certainty. However, it is logical that fewer people in the water would inherently lessen the risk of shark attack.

A spate of shark bites in the US during the summer of 2001, and the incessant media reporting (and re-reporting) of those incidents, led to this period being termed 'Summer of the Shark'. In response, a state-wide legislated ban on provisioned fish-feeding was implemented throughout Florida waters on 1 January 2002 (68B-5.005 Divers: Fish Feeding Prohibited; Prohibition of Fish Feeding for Hire; Definitions). In context, the term 'provisioning' was defined as an activity that offers some type of attractant, bait or food reward to aggregate or positively reinforce wild animals and/or neutralise their aversion towards humans for the purpose of tourism. The practice of provisioned feeding of marine animals is not uncommon, and is widely employed to attract these often-elusive animals to tourism operations. Provisioning wildlife for tourism has become increasingly popular with the growing demand for people to closely observe and interact with animals in their natural settings, and shark feeding has been identified as the 'most developed provisioning activity' internationally (Clua *et al.* 2010). A 2013 estimate suggested that the shark diving tourism industry attracts over half a million participants annually over ~85 countries (Cisneros-Montemayor *et al.* 2013), and no doubt this figure has grown. Wildlife provisioning practices are variable depending on target species, cultures and regulations, and result in positive and negative impacts on both the animals and the people involved. The practices are highly controversial for several reasons, and are banned in many marine protected areas. Long-term, targeted food provisioning to wildlife has been shown to modify natural behavioural patterns and population levels, cause dependency on human-provisioned food and habituation to human contact, have dietary impacts on the animals, disrupt natural species assemblages and association, and lead to intra- and interspecific aggression, including harm to the people involved (Orams 2002; Burgin and Hardiman 2015). Bans on provisioning sharks are in place in some regions in the US, the Cayman Islands and South Africa. Although applicable to fish in general, the Florida regulations were largely directed at local commercial dive operators, who were blamed by some for increasing shark aggression towards spearfishers and other water users by conditioning sharks to associate humans with food.

The Florida Government's ban on fish provisioning has been considered a simple placebo response in shark attack mitigation (Sunstein and Zeckhauser 2008). But although potentially seen as petty, in 2014, charges were filed by the Florida Fish and Wildlife Conservation Commission against four people connected to the illegal feeding of sharks within state waters during dive charter trips. The charges were in response to complaints made by members of the public who had witnessed the events and had suggested that the feeding activity resulted in the sharks becoming extremely aggressive towards humans who were in the water watching the activity. All four offenders were charged with operating a vessel that conducted fish feeding for the value of passengers within state waters, and two were additionally charged with fish feeding. These were second-degree misdemeanours, punishable by up to 60 days in jail and a fine of up to US$500.

A ban on shark feeding in Hawaiian waters (Code [§188–40.6] Shark feeding; prohibitions; exceptions; penalties) was implemented in 2010, the rationale being 'the practice of feeding sharks is not in the interests of the people of Hawaii. This activity is exploitative and concentrated shark populations are a problem for Hawaii's ocean recreation community. An increased shark population in well-known surfing, diving, fishing, and other marine recreational areas increases the probability of shark attacks' (State of Hawaii 2010). The Western Australian Government implemented legislation in 2012 banning targeted or dedicated shark tourism ventures that relied on attracting

Sharks have been revered and respected in Hawaii for as long as humans have made their homes on these islands. But the ocean is the shark's house, and people who enter it accept a certain level of risk. The chances of being bitten by a shark in Hawaiian waters are extremely small; injuries range from minor to severe, occasionally even fatal. So it's important to recognise the risk, and take steps to minimize it. Enjoy the ocean, but remember you're just a visitor in a world that belongs to the sharks.

Suzanne Case
Chair, Hawaii Department of Land and Natural Resources

sharks, including cage diving, and proposed bans on the use of offal and blood to fish for sharks in close relation to popular beaches (DoF 2012b).

Shark bite prevalence is still following an upward trend in Florida, despite the 2002 ban on fish feeding. Based on data from the Global Shark Attack File, in the 15 years since 2002 there have been an average of 24.7 unprovoked attacks per year in the state; in the 15 years before the ban, there were an average of 15.3 unprovoked attacks per year. However, it must be acknowledged that many other factors may also affect shark–human interaction and would have contributed to bite prevalence rates during those periods.

Effectiveness of legislative bans on feeding and interacting with sharks in mitigating the risk of shark attack

Legislated bans related to mitigating the risk of shark attack are often highly unpopular with the local community. Media reports following the implementation of such bans attract many comments suggesting that people need only be educated and informed, then allowed to make their own decisions about entering the water, or that the bans are misdirected or ill-conceived. Not surprisingly, there are always people who continue to engage in the banned activity despite the legislation. A commercial dive operator in Florida (described above) continued to feed fish and sharks, and surfers and swimmers continued to engage in water-based activities at banned beaches in Brazil and Réunion Island (sometimes with fatal consequences) due to disregard for the bans, or because of a lack of knowledge or understanding.

The real question is whether such bans, when successfully employed, are effective in mitigating the risk of shark attack. While full bans on water usage would of course be completely effective in mitigating the risk of shark attack if they are embraced and upheld by the human population, they are clearly not a realistic option for use over a protracted period. The effectiveness of bans on feeding, attracting and interacting with sharks remain controversial in terms of shark attack mitigation.

Provision-induced behavioural changes have been studied under various conditions and locations, and among different shark species. Chumming (baiting) the water in Australia and South Africa to attract white sharks (*Carcharodon carcharias*) for cage-diving tourism has led to significant increases in shark abundance, residency periods and changed diel patterns. However, changes were limited to a portion of the sharks in the immediate vicinity, and no changes were observed in sharks only 12 km from provisioned sites (Johnson and Kock 2006; Bruce and Bradford 2013). Habituation was limited to the months that provisioning occurred, and the time sharks spent at the boat and their interest in the boat and the provisions lessened over time. Provisioning has been also shown to affect the species composition, number of sharks and residency patterns in other areas (Clua *et al.* 2010; Fitzpatrick *et al.* 2011; Maljković and Côté 2011). In Hawaii, shark provisioning for cage-diving altered the assemblage composition of numerically dominant sharks over a five-year period, with the abundance of relatively small (less dangerous) sandbar sharks (*Carcharhinus plumbeus*) declining, while larger species such as Galapagos sharks (*Carcharhinus galapagensis*) and tiger sharks (*Galeocerdo cuvier*) increased (Meyer *et al.* 2009b). Multiyear studies of up to eight species in provisioned areas in Fiji showed that even though bull sharks (*Carcharhinus leucas*) were not hand-fed, their numbers increased and they became the most numerous species in the area, their site fidelity was altered to reflect the provisioning times (even on non-provisioned days), they stayed at the provisioning location for longer periods and their overall diel movement patterns changed (Brunnschweiler and Baensch 2011; Brunnschweiler and Barnett 2013; Brunnschweiler *et al.* 2014).

A provisioned whitetip reef shark (*Triaenodon obesus*). Photo by Tyrone Canning.

Perhaps the most damning piece of evidence linking shark aggression to human provisioning is the report by Levine *et al.* (2014), which found that in a spate of five unprovoked shark attacks on humans in Sharm-El-Sheikh (Egypt) between 30 November and 5 December 2010, three were attributed to an individual oceanic whitetip shark (*Carcharhinus longimanus*) identified by a distinctive crescent-shaped notch on its caudal fin. In each of the three cases, the shark was observed and identified by a witness or photograph. Video footage taken several months before the attacks showed the same shark being hand-fed underwater by a diver, and swimming behind the diver while he was taking another fish out of a bag located over his buttock. Even more revealing is that each victim lost a hand and/or a portion of their forearm and that, during at least two of the attacks, an extensive amount of tissue was removed from the victim's buttock – the exact places from which food had previously been sourced for the shark. While this is highly suggestive of a correlated link, other results are fragmented and not uniform across interactive activities and the species involved. Even in the case of the oceanic whitetip, the same shark was observed and photographed two days after the final (fatal) incident, interacting with divers without any negative results.

Whenever there is change in the residency pattern of sharks, and when even part of the population becomes habituated to provisioned sites (which are generally sites where tourism activities are commonly carried out), there is an increased potential for shark–human interaction. However, in certain circumstances where the potential for negative shark–human interaction is low, the capacity for humans to observe sharks up close in their natural environment could help to lessen the human fear of sharks through enhanced understanding of the animals and non-negative interactions. This could also help to mitigate the impression that all shark encounters end negatively. But ultimately, the crucial question is whether shark diving operations and/or provisioning of sharks for tourism does present a significant safety concern for water users through habituation of potentially dangerous species to humans or a learnt association of humans with a food reward. Unfortunately, a definitive answer regarding sharks remains elusive. Various studies suggest that it is unlikely that provisioned cage-diving activities (at least) present a risk for increased attacks on humans in nearby regions (Johnson and Kock 2006; Laroche *et al.* 2007; Meyer *et al.* 2009b). However, if a negative response was conditioned, it could prove disastrous. Thus, a precautionary approach to such activities has been proposed for operators and tourists, namely to prohibit operation in close proximity to common human use areas, direct interaction with the animals through touch, hand-feeding, manipulating or harassment, or interference with natural shark behaviours (Orams 2002; Gallagher *et al.* 2015). It is also imperative that provisioning locations be openly identified and broadcasted, so that other water users are aware of this sort of activity and the chance of increased, and potentially altered, shark presence and behaviour. The provision of bait that is significantly different from normal prey, and given only in very small quantities, may also help to separate provisioning from actual feeding events (Clua *et al.* 2010).

Conclusion

In most cases, legislation relating to the threat of shark attacks is futile. However, in some places, governments do need to address the issue. In those instances, what is needed is considered, proactive legislation that responds to scientifically and statistically backed fact, not to emotion.

Certain human-initiated activities, such as provisioning, may increase the likelihood of negative shark–human interaction and such activities should be reconsidered. Most importantly, communication is key. This relates to open, timely and continued government communication with the public regarding new legislation or policy. The background logic that contributed to such decisions should be clearly explained. Open and honest communication to the public needs to be mandatory if sharks are being attracted to certain regions or to humans for any purpose (e.g. provisioning for tourism activities, media filming, fishing competitions), and the people responsible for those activities should be legally obliged to produce that communication. Effective, appropriate and continued communication will maximise the success of legislated risk mitigation.

Due to the nature of the risk, governance in relation to shark attack must be flexible and incorporate ongoing monitoring and assessment procedures. Legislation and policies must be able to be modified based on reviews of their efficacy in terms of actual and perceived shark attack mitigation, and biodiversity outcomes. As scientific research continues to provide additional information on sharks, as well as the interactions between sharks and humans, that information should be taken into account and programs altered accordingly. And finally, educational programs and resource outlays need to be considered and incorporated into sustainable regimes.

11

Looking towards the future

Even though shark attack incidence is increasing globally, we have come a long way in our understanding of these events. The main reason for the rise is simply the increasing amount of time humans spend in the water. That is a given, and completely predictable. However, we're also beginning to zero in on other factors that contribute to increased prevalence, which will allow for the development of more appropriate mitigation measures. Although we have not succeeded yet in producing fully effective personal or regional shark attack mitigation devices, we are more knowledgeable about them than we were 60 years ago, when the 'best' shark deterrent devices would not only *not* deter sharks from biting but also killed a variety of animals in the immediate vicinity. We know so much more about shark behaviour, movement patterns and sensory biology – all of which allow us to construct more effective mitigation strategies. As we continue to learn about sharks, new strategies and devices will be designed and refined, and it is fully expected that these measures will help to reduce shark attack incidence around the world.

This may be going out on a limb, but I also believe that many of us are finally starting to forgive and forget *Jaws*. This is happening, surprisingly, in conjunction with (but possibly also due to) shark attacks becoming regular and mainstream in the media. I believe that, on the whole, our attitudes about sharks are shifting from fear and hatred to empathy and conservation. There are many reasons why we need to conserve the sharks in our oceans, including those we currently realise and others that we are yet to learn. Pressure is being put on politicians to adopt non-lethal mitigation measures (where measures are required at all), and the implementation of lethal measures is often vociferously challenged even in the wake of spates of shark attacks. Although rarely required, shark attack legislation and policy is sometimes warranted. To maximise the effectiveness of outcomes and to most appropriately meet the desires of the general public, legislation should be proactive and constructed from a multidisciplinary approach. Decisions should be made through collaborative discussions between government officials, researchers, conservation groups, industry bodies and representatives of the general public. The media should be brought on board early, and kept informed and educated on developments. Through this, all stakeholders will feel involved in the decision-making process and will be more willing to support outcomes, providing a more unified front against the potential risk sharks may pose.

Studies suggest that the measures most commonly desired by the general public for alleviating the risk of shark attacks are better education strategies and advanced warning systems. The majority of people do not wish any harm to sharks, they just want to be better informed in order to make their own decisions on how they can safely coexist with sharks. Interestingly, these preferred measures largely place the onus and accountability of risk mitigation onto individual water users, although they do require government initiatives for monitoring shark movements and educational campaigns. Ideally, education on understanding, respecting and coexisting as safely as possible with sharks should begin at an early age. Understanding and accepting the fact that sharks are natural inhabitants of open waterways, and hence there is a risk to humans who also use those environments, is also necessary.

Although it can't be stressed enough that the incidence rate of shark bites is incredibly low and that many incidents result in only minor trauma to the person, major shark attacks and fatalities do occur. As a result, we must be at the forefront not only of shark attack mitigation policy and research, but also in treating and assisting those who have been affected. While we have made great progress in optimising life-saving first aid and emergency trauma medicine, which has led to a significant decrease in shark-related fatalities, we need to be much better at providing post-surgical mental and physical support for people affected by major shark attack trauma. Major attacks can be incredibly isolating experiences, and an overall lack of support can present additional challenges for recovery. While prevention is always the best policy, this does not mean that we should not dedicate resources to assist those who have been affected. Beneficial resources may take the form of post-operative medical or trauma treatment, or informational resources detailing support providers or organisations. Those who have been bitten would obviously need the most support, but we must remember that these events have the potential to affect a wide range of others who may also need support.

Conclusion

Few species are more feared than sharks and quips have been made stating: 'If they could sue, [sharks] could make a case for defamation of character', and that they are 'a victim of poor public relations' and 'doomed to play the villain' (Papson 1992). However, sharks suffer from poor public image only partly because they potentially threaten human safety. We are subconsciously biased towards animals that are most similar to us and prefer those that are tame, display human-like characteristics and exhibit behaviours we can understand. We tend to dislike those we consider to be wild, dangerous, unpredictable, alien in adult appearance, that aren't cute even as babies, that use a prolonged or gruesome method of killing, and that use senses that we cannot comprehend or feel (Burghardt and Herzog 1989; Woods 2000). However, people are often intrigued by danger and difference (Ryan 1988). One study of Australians showed this polarisation, finding that sharks were the least favourite animal for 16% of respondents, but the favourite of 9% (Woods 2000). People ranked animals based more on their perceived attributes than on their actual attributes. The negative perception of sharks and the risk of shark attack has been recognised as not only a social and political challenge, but also a barrier to global shark conservation.

Despite the seeming opposition, shark attack mitigation and shark conservation are linked: shark bite prevention is shark conservation (Neff 2014). It is not only possible to show empathy to both sharks and humans that have negative interactions with sharks, it has also been shown that this is the most common feeling. Unprovoked shark attacks are nothing more than highly unfortunate accidents, and blame for such accidents cannot, and should not, be attributed to either party. Discussions of the endangered nature of sharks and the lethality of shark attack mitigation measures immediately following shark attacks may not only further traumatise those affected by the event, but may hinder conservation efforts. It would therefore be in the best interest of conservation groups to acknowledge the trauma of these situations and offer condolences to everyone involved, to be seen as proactive in the preservation of all life. This is particularly relevant during highly emotive times, especially following serious or fatal bites, when the destruction and removal of sharks from the environment may be more welcomed by some members of the public. Conservation efforts need to support and encourage appropriate shark attack mitigation strategies and work with other stakeholders to identify optimal outcomes.

I was bitten in 1997. When it happened, it was really vivid, but I've spoken about it so many times now since then that I have started to wonder if over time, I've begun to think of it a bit differently. But I remember it all really clearly and I can still picture the attack in my head.

I was born in Hawaii and grew up surrounded by the ocean. I really loved body-boarding, and always wanted to be a professional body-boarder; it was my destiny. In the wintertime, we get really good swells in Hawaii. This is in complete contrast to summertime, where the surf is generally really small. So by September, my body-boarding team was just holding out for really good waves. I had just graduated high school, and was in that transition period of trying to turn professional and get a paid living as a body-boarder. As the ocean started to get more and more churned up, we got more and more excited for the waves. But our first good swell of the year didn't actually come until late October.

We all woke up really early that morning and went down to the beach. When we got there, I remember being overwhelmed by a really stinky smell, like a foul, dead fish smell. I had never smelled anything like it before in my life, it was like you walked into a fish auction block, in a closed room, after the air-conditioner had been off for a month. But we were just aching to get into the water, so we made the decision to go in anyway. I remember walking into the water with one of my teammates, probably around 7:00 or 7:10 in the morning. We both jumped into the water at the same time, and he bumped into my shoulder. I got this really weird feeling, like a creepy, get away from me, fight or flight feeling. I didn't think that it was a shark or anything unusual, but I just got such a weird feeling from what would have otherwise been a completely insignificant interaction. I remember thinking how weird that feeling was.

We paddled out and got into a line-up, which had me ~50 yards off the beach, in ~30 feet of water. The visibility was pretty average. Everybody on my team caught rides into the beach, but I held back, really just wanting that perfect wave. So by the time the last wave of the set came in, I had been sitting there completely motionless on my board in a vertical position with my legs dangling over the edge and my swim fins not moving at all for maybe a good six minutes or so. The only other person out there with me was a surfer that I had never seen before. I remember he looked at me, and I looked back at him, and it was like, who is going to go for this wave. You don't ride waves together, it's just bad etiquette. But as soon as you make that initial stroke, that's telling the other person that the wave is yours. And I distinctly remember making that first stroke.

My limbs had been completely static and motionless, but as soon as my fingertips hit the water, breaking the surface, a large tiger shark came up vertically, like a submarine, right out of the water and grabbed onto my leg. And instantly it was that same feeling that I got jumping into the water where I bumped my teammate. But this time, it was obvious that I was getting attacked by a shark. The shark was right at my chest, and I was looking right at it. It just felt like this crazy, enormous pressure, like the entire team of guys were all sitting on my leg at the same time.

There was no pain whatsoever, but the pressure just kept getting stronger and stronger. The shark was obviously trying to clamp down on my legs. I distinctly remember sticking my right hand in its mouth to try to prevent it from biting down on my leg, but that didn't work. Its nose continued to press into my chest, and it carried out this real violent left-and-right shaking, like a dog would do with some meat. Then it stopped. And I remember it opening its jaws again, like it was trying to get a better grip on my leg, and that's when I took my left hand and punched it, just totally out of instinct. You know, like 'fight or flight, let me try this', and I hit the shark. On maybe my third hit, it let go and swam away. It was all one attack, and it would all have taken place over about five seconds. But in that time, it got my hand, my left leg and my right leg.

I got back on my board, and I looked at the other guy who was maybe 5 feet away, and he was really pale. I yelled, 'Shark, go in!' and he took off paddling. I looked at my index finger that I had stuck in the shark's mouth, and it was split open like a potato. I could see my bone and blood and guts and it was pretty gross. I had never seen that much blood in my life. And at that point, I was like, 'Oh, I'm hurt', and then started paddling right behind the other guy back towards the beach.

I specifically remember this crazy, uncontrollable shake in my right leg, like a completely uncontrollable spasm. You don't see your legs from the paddling position, your legs are behind you, and I hadn't taken the time to look back. So when the shaking began, I thought it was the shark on my leg again, coming to finish me off. But when I looked over my shoulder, it wasn't the shark, but my leg. It was perfectly amputated off, as clean-cut as you could imagine, as if someone had taken a chop saw or an axe and cut it right off. And it was squirting blood every time my heart beat. It was at that stage that I had an, 'Oh, no, I could die at any moment' panic kind of moment.

Luckily, a second or two later, I caught a little wave all the way up to the sand. When I got onto the sand, I tried standing up, but you can't stand up without a foot. I fell over and rolled back down the sand into the ocean. My friends, who had all seen me coming in, knew that something was wrong when they saw the blood. They grabbed me and dragged me up the beach. One of my teammates immediately took my leash and made a tourniquet around my thigh. It was very smart, very clear thinking, and later I learned that that action had saved my life. The same guy then started saying a prayer for me. I had had a real scary 'I'm going to die' feeling from the moment I saw my leg missing. But I remember closing my eyes when he started saying the prayer, and I felt a real calm come over me as everyone was praying for me.

When I opened my eyes, there was a truck there. I don't know how it had gotten there so fast, but I guess they had seen the commotion. The guys lifted me up into the bed of the truck, and the guy who was driving drove as fast as he could on the sand and then onto the road to get me to the hospital. As it happened so fast, nobody thought to jump in the back of the truck with me, or anything, so I was there by myself, with the tailgate still down. There were some surfboards in the back with covers, woven cloth bags that protect them from the sun's UV rays, and I remember not wanting to look at the injury, so I instead just focused on this fabric. And I remember getting really hot, then really cold, then really hot, and cold again.

At one point, I looked back through the tailgate, without looking at my leg, as we passed a car. It was a mother taking her daughter to school, and I remember seeing the look on their faces. At first they looked so mad, probably because we passed them so fast, and then it was this look of total shock and I could see it in their eyes. You can always tell the extent of an injury by the reaction of others. Then all of a sudden, we pulled up to a small little hospital called Kauai Veterans Memorial Hospital in Waimea. We pulled in so fast, right in front of the emergency room doors. The doors swung open, and a nurse and a doctor ran straight out to me. The doctor put his hand on my shoulder, and I didn't completely lose consciousness but went into this weird dream-like state. When I came to, probably about half an hour later, I was naked, and being hooked up to all these needles and IVs and was being given transfusions and things. I remember telling the people in the room how big the shark was, and how it had a real square nose, and trying to give them the length, but then I passed out again. When I woke again, I was in an ambulance heading towards our main hospital. My dad is a paramedic, and it was one of his co-workers in the ambulance with me and he started telling me all of these shark jokes. And even in my dream state, I remember laughing at these jokes, and then I passed out completely.

It must have been 24 hours later when I woke up again, and I was in our main hospital, Wilcox Memorial, and my family was all around me. The doctor was

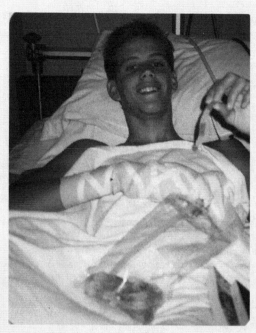

Mike Coots in the hospital after being attacked by a tiger shark (*Galeocerdo cuvier*) in Hawaii. Photo supplied by Mike Coots.

there too, as well as my teammates. They all looked really sad, and my mom and my dad looked at each other, and the doctor looked at them, and they were all saying, 'Are you going to tell him?' Finally, my mom told me I had lost my leg. I said, 'I know, I was there when it came off', and everyone started laughing. And I think that once they realised that I knew that I was going to be an amputee, then they were all just happy knowing that I was alive, and everyone had more of an uplifting moment.

I was in the hospital for about a week getting everything stitched and stapled up. And then I was back home and in and out of rehabilitation to get my muscles working again. It was another three or three and a half weeks until the stitches and staples came out. But as soon as they did, I was back in the water body-boarding. It was really hard, that three and a half weeks of being out of the water, because it was the very beginning of the season and we had had a bad summer, so I just couldn't wait to get back into the water. That was the hardest part of the attack! It wasn't losing the leg, it was just being out of the water that long.

The shark (which the experts later told me was probably ~12–15 feet) took my right leg off a bit above my ankle, and the doctors took a bit more off to make it an optimal residual length to fit a prosthetic, but also to make sure the site was totally clean because sharks teeth can be dirty and have a lot of bacteria. The shark also got my left foot. I actually think that I lost more blood from that foot, and I have some pretty big scars to show for it. And it got my right index finger when I stuck my hand in its mouth, and I had to get nerve surgery on that finger. My finger had no sensitivity for five or six years or maybe even longer, it was just completely numb. The doctors later told me that with the amount of blood that I lost, it was a medical miracle that I survived. With most shark attacks where you lose a limb, there is a good chance that you're not going to make it, and I additionally had that trauma on my left foot that bled quite a bit too.

I had no hesitation about getting back in the water after the attack. Growing up, I knew nothing else, the ocean was everything. And it just so happened that when I got back in the water, the waves were good at the spot that I got attacked, so I actually went back in for the first time really close. Not the exact same wave, but the same stretch of span where I had been attacked. I wasn't one bit scared that it would happen again. I mean, statistically, the chance of that happening again would be astronomical, and I could write a pretty good book myself if it were ever to happen again. I think had I been a tourist or a visitor here from the mainland on holidays, having not spent too much time in the ocean previously, then it would have been a different story. But because I had

Mike coming face to face with a large adult tiger shark (*Galeocerdo cuvier*) for the first time since his attack 20 years ago. Photo by Mike Coots.

spent basically every second of my free time since I was about six years old in the ocean, it was such a freak, weird thing, and I just don't think that it will ever happen again. But if it does, then so be it, I spend a lot of time in the water. I also had no trouble getting back into body-boarding, because with body-boarding you just hang your legs off. I wear one fin, and it actually works in my favour, because I have less drag from an extra leg and fin behind me, but I still have one fin to use to catch waves. So it actually helps me and I have a little bit of an advantage and I can go a bit faster.

I've been very fortunate, I've never had one bad dream or any flashbacks. In fact, I still dream that I have two legs, it's really weird. I've just been to Tiger Beach in the Bahamas, and it was my first time coming face to face with tiger sharks since the attack nearly 20 years ago. I spent a week diving, and it was really amazing to be able to dive every day with large tiger sharks. I've done a lot of diving, and I've been in the water with pretty big great whites, and I fed a tiger shark from the surface once, and that was pretty cool. But actually being underwater, looking eye to eye with large tiger sharks, was amazing. I figure if the flashbacks were going to come, they would have done so with a huge tiger shark right in my face, but they never did! I do still have very specific memories of the event, though. The first thing that really stands out in my mind was that smell, definitely that smell, unlike anything I had ever experienced before. Second, the very squareness of the shark's nose. It would have been completely 90°, really square and really wide. Third, the feeling that I got. Normally I only get that feeling around centipedes, which are really the only scary thing we have here in Hawaii, other than sharks. But it was like when you see a snake, and the hairs on you stand up, and you just get that feeling. And I had that feeling twice within 10 minutes, when normally I go years without feeling like that.

I imagine that the shark that bit me was in the area because of whatever was causing that smell, and I probably looked like a turtle. It was probably underneath me, checking me out because I had been motionless for so long. Then as soon as I made a movement, with a splash and the energy of movement, it attacked. I've got a couple of theories about why the water smelled like it did that day. They are all anthropogenically related (stemming from poor aquaculture or fish-farming practices). But as none of these theories can be proven, there is no point in laying blame, and I haven't pursued my investigations any further. And the way I see it, everything in my life has worked out really well, so I've got no resentment. But since I was bitten, at least one other person has also been bitten under similar conditions. The description that that person provided of the event revolved around that same characteristic overwhelming smell of rotting fish.

I visit and speak to other shark attack victims as much as possible right after their attacks. In some cases, I have been the only one there when a person's come out of surgery following an attack. It can be pretty surreal. I even went to Boston after the Boston Marathon bombing in 2013 to talk to the victims of the blasts there in the hospital. I think it can be a relief to have someone who has also been attacked by a shark and lost a limb to be right by your side from really early on. I'm a strong advocate that that's one of the most important things that can be done for optimal recovery. And it's sort of like a special club that you're in. I hope that other shark attack survivors also do this, or other people who have lost limbs for whatever reason. How to feel after the loss of a limb from a shark is not the type of thing that you can ask your doctor or your nurse or your family members about. It's sort of a personal thing. And when I was in the hospital, I had no idea about prosthetics. But about three days into my week-long hospital stay, I had a guy visit me. I had no idea who he was, and he was wearing pants, but he chatted to me for five or 10 minutes and when he left my mother told me he had a prosthetic leg. And that was when it dawned on me that I would be able to walk again. And it really helped me early on to know that there were other people out there like me. I think that it's kind of why we're here on Earth, if there is some unique way that we can help others, or directly help someone else, then that's our calling. And if you don't do things like that, then I don't know how you can sleep at night, especially after other people have been there for you. In relation to shark attacks, I think right now people just take it upon themselves to reach out to others in their regional areas. But everything is just so connected now, that within about five minutes of hearing about an attack, I have been able to get email addresses and phone numbers of the person. I think that that's the best thing about social media and the community – the ability to be able to connect with other people in need. I feel fortunate that I can help others.

None of my family, friends or I have any hatred or hard feelings towards sharks, it's just one of those things that if you spend enough time around an island surrounded by the ocean, it just goes with the territory. From my experience, there is a 100% feeling among surfers and body-boarders that sharks are just a part of the activity. And it's a bit of a yin and a yang; the ocean gives us so much joy, happiness and great memories, and sharks really just make the ocean even more interesting. It's pretty cool to essentially have dinosaurs swimming in the ocean and they help to give the ocean that mysterious, dangerous feel. If there were no sharks, the ocean would be a very different place and it wouldn't instil that curiosity or healthy fear. Everyone I know has a really healthy fear of sharks, and spearfishers even more so. They're both sort of hunting fish at the same time. I don't know anyone that thinks sharks are bad, or should be removed from our ecosystem, or anything like that.

Culling is obviously the worst form of shark attack mitigation, and I'm very hopeful for some of the new technology that is being developed. I heard stories that there were a lot of fishers that went out to see if they could catch the shark that bit me right after my attack. But in hindsight, I am glad that no official retaliatory action was taken, as was done following some of the other shark attacks in Hawaii. I think right now our best bet for mitigation is a combination of measures. Réunion Island has been having a lot of trouble with attacks, and they've instigated a team of local shark spotters who go out in scuba gear from a dive boat to the areas that are popular surfing areas and spot for bull sharks. They

Mike surfing with his prosthetic leg. Photo supplied by Mike Coots.

have a flagging system for if a shark comes into the area, in which case the surfers will leave. It creates jobs, keeps the surfers safe and it isn't harming the animals. There are obviously nets, which have a downside, but they're better than culling. I think everything except culling has advantages. I've been approached by various companies to endorse personal mitigation products, but I haven't gone through with any of them. I think that if there was something that really worked, it would be a billion-dollar product and there would be a lot of people around the world using it. But I think right now, a lot of the products on the market have the potential to attract sharks. If you're sending out some sort of signal, then you're sending out something that might pique their curiosity. I really believe in just trusting your instincts and, if something doesn't feel right, not going in the water. I think the acoustic monitoring systems that send alerts about shark presence are wonderful and I think that that's a great way for easing a lot of people's worries, and that's using science to allow us to better coexist. But if I was alerted to the fact that there was a large shark in the water, would I still get in the water? I guess it depends on how good the waves are! And maybe to an extent what the other conditions are. If it's pouring rain and poor water visibility and you're in a river mouth and there is an alarm pinging that there is a large tiger shark in the area, then getting in the water would probably be pretty irresponsible. But sharks swim pretty fast, so they can cover a great distance over a very short period of time. The alert is just one of those things that you'd have available for additional information to make your own choice, depending on how accurate you think the information is. But ultimately, the ocean is just too much fun to worry about something like sharks.

Mike Coots

References

Achen CH, Bartels LM (2004) Blind retrospection electoral responses to drought, flu, and shark attacks. *Estudio/Working Paper 2004/199*. Center for Advanced Study in the Social Sciences, Juan March Institute.

Adams DH, Borucinska JD, Maillett K, Whitburn K, Sander TE (2015) Mortality due to a retained circle hook in a longfin mako shark *Isurus paucus* (Guitart-Manday). *Journal of Fish Diseases* **38**, 621–628. doi:10.1111/jfd.12277

Adlington K, El Harfi J, Li JN, Carmichael K, Guderian JA, Fox CB, Irvine DJ (2016) Molecular design of squalene/squalane countertypes via the controlled oligomerization of isoprene and evaluation of vaccine adjuvant applications. *Biomacromolecules* **17**, 165–172. doi:10.1021/acs.biomac.5b01285

Allen K (2015) Comment on 'Aquatic animal telemetry: a panoramic window into the underwater world' by Hussey NE *et al.*, *Science* **348**. http://comments.sciencemag.org/content/10.1126/science.1255642#comment

Amin R, Ritter E, Wetzel A (2015) An estimation of shark-attack risk for the North and South Carolina coastline. *Journal of Coastal Research* **31**, 1253–1259. doi:10.2112/JCOASTRES-D-14-00027.1

Amorim A, Arfelli C, Fagundes L (1998) Pelagic elasmobranchs caught by longliners off southern Brazil during 1974–97: an overview. *Marine and Freshwater Research* **49**, 621–632. doi:10.1071/MF97111

Aronson LR, Gilbert PA (1958) Conference on shark repellents. *A.I.B.S. Bulletin* **8**, 17–19. doi:10.2307/1292385

Arrindell WA, Pickersgill MJ, Merckelbach H, Ardon MA, Cornet FC (1991) Phobic dimensions: III. Factor analytic approaches to the study of common phobic fears: an updated review of findings obtained with adult subjects. *Advances in Behaviour Research and Therapy* **13**, 73–130. doi:10.1016/0146-6402(91)90014-2

ASAF (2016) Australian Shark Attack File. Taronga Conservation Society Australia. Accessed 20 October 2016.

Atkinson CJL, Collin SP (2012) Structure and topographic distribution of oral denticles in elasmobranch fishes. *Biological Bulletin* **222**, 26–34. doi:10.1086/BBLv222n1p26

Baldridge HD (1974) Shark attack: a program of data reduction and analysis. *Contributions from the Mote Marine Laboratory*. 2nd edn. Sarasota.

Baldridge HD (1988) Shark aggression against man: beginnings of an understanding. *California Fish and Game* **74**, 208–217.

Baldridge HD Jr (1990) Shark repellent: not yet, maybe never. *Military Medicine* **155**, 358–361.

Baldridge HD Jr, Williams J (1969) Shark attack: feeding or fighting? *Military Medicine* **134**, 130–133.

Banks N (2015) Submissions of Sea Shepherd Australia Ltd: Management of Sharks in New South Wales Waters (Inquiry). Melbourne, Sea Shepherd Australia.

Barker WH, Weaver RE, Morris GK, Martin WT (1974) Epidemiology of *Vibrio parahaemolyticus* infection in humans. In D Schlessinger (ed.), *Microbiology – 1974*. Washington, DC, American Society for Microbiology.

Baum JK, Myers RA (2004) Shifting baselines and the decline of pelagic sharks in the Gulf of Mexico. *Ecology Letters* **7**, 135–145. doi:10.1111/j.1461-0248.2003.00564.x

Baum JK, Myers RA, Kehler DG, Worm B, Harley SJ, Doherty PA (2003) Collapse and conservation of shark populations in the Northwest Atlantic. *Science* **299**, 389–392. doi:10.1126/science.1079777

Beck U (1992) *Risk Society: Towards a New Modernity.* (English translation). London, Sage Publications.

Bernal D, Sepulveda C, Graham JB (2001) Water-tunnel studies of heat balance in swimming mako sharks. *Journal of Experimental Biology* **204**, 4043–4054.

Bonfil R, Meÿer M, Scholl MC, Johnson R, O'Brien S, Oosthuizen H, Swanson S, Kotze D, Paterson M (2005) Transoceanic migration, spatial dynamics, and population linkages of white sharks. *Science* **310**, 100–103. doi:10.1126/science.1114898

Brenneka A (2009) 1992/04/24 Mike Fraser – New Zealand – Campbell Island. *Shark Attack and Related Incident News Archive.* Shark Attack Survivors. http://sharkattacksurvivors.com/shark_attack/viewtopic.php?t=1167

Brodie P, Beck B (1983) Predation by sharks on the grey seal (*Halichoerus grypus*) in eastern Canada. *Canadian Journal of Fisheries and Aquatic Sciences* **40**, 267–271. doi:10.1139/f83-040

Brown AC, Lee DE, Bradley RW, Anderson S (2010) Dynamics of white shark predation on pinnipeds in California: effects of prey abundance. *Copeia* **2010**(2), 232–238. doi:10.1643/CE-08-012

Bruce BD, Bradford RW (2013) The effects of shark cage-diving operations on the behaviour and movements of white sharks, *Carcharodon carcharias*, at the Neptune Islands, South Australia. *Marine Biology* **160**, 889–907. doi:10.1007/s00227-012-2142-z

Bruce B, Bradford R (2015) Segregation or aggregation? Sex-specific patterns in the seasonal occurrence of white sharks *Carcharodon carcharias* at the

Neptune Islands, South Australia. *Journal of Fish Biology* **87**, 1355–1370. doi:10.1111/jfb.12827

Bruce B, Stevens J, Malcolm H (2006) Movements and swimming behaviour of white sharks (*Carcharodon carcharias*) in Australian waters. *Marine Biology* **150**, 161–172.

Brunnschweiler JM, Baensch H (2011) Seasonal and long-term changes in relative abundance of bull sharks from a tourist shark-feeding site in Fiji. *PLoS One* **6**, e16597. doi:10.1371/journal.pone.0016597

Brunnschweiler JM, Barnett A (2013) Opportunistic visitors: long-term behavioural response of bull sharks to food provisioning in Fiji. *PLoS One* **8**, e58522. doi:10.1371/journal.pone.0058522

Brunnschweiler JM, Queiroz N, Sims DW (2010) Oceans apart? Short-term movements and behaviour of adult bull sharks *Carcharhinus leucas* in Atlantic and Pacific oceans determined from pop-off satellite archival tagging. *Journal of Fish Biology* **77**, 1343–1358. doi:10.1111/j.1095-8649.2010.02757.x

Brunnschweiler JM, Abrantes KG, Barnett A (2014) Long-term changes in species composition and relative abundances of sharks at a provisioning site. *PLoS One* **9**, e86682. doi:10.1371/journal.pone.0086682

Bryson M, Williams S (2015) *Review of Unmanned Aerial Systems (UAS) for Marine Surveys.* University of Sydney, Australian Centre for Field Robotics.

Buck JD, Spotte S, Gadbaw JJ (1984) Bacteriology of the teeth from a great white shark: potential medical implications for shark bite victims. *Journal of Clinical Microbiology* **20**, 849–851.

Burgess GH, Callahan MT, Howard RJ (1997) Sharks, alligators, barracudas, and other biting animals in Florida waters. *Journal of the Florida Medical Association* **84**, 428–432.

Burgess GH, Buch RH, Carvalho F, Garner BA, Walker CJ (2010) Factors contributing to shark attacks on humans: a Volusia county, Florida, case study. In JC Carrier, JA Musick, MR Heithaus (eds), *Sharks and their Relatives II: Biodiversity, Adaptive Physiology and Conservation.* Boca Raton, CRC Press.

Burghardt G, Herzog H Jr (1989) Animals, evolution, and ethics. In RJ Hoage (ed.), *Perceptions of Animals in American Culture.* Washington, DC, Smithsonian Institution.

Burgin S, Hardiman N (2015) Effects of non-consumptive wildlife-oriented tourism on marine species and prospects for their sustainable management. *Journal of Environmental Management* **151**, 210–220. doi:10.1016/j.jenvman.2014.12.018

Burnett JW (1998) Aquatic adversaries: shark bites. *Cutis* **61**, 317–318.

Bushman BJ (1998) Priming effects of media violence on the accessibility of aggressive constructs in memory. *Personality and Social Psychology Bulletin* **24**, 537–545. doi:10.1177/0146167298245009

Byard RW, Gilbert JD, Brown K (2000) Pathologic features of fatal shark attacks. *American Journal of Forensic Medicine and Pathology* **21**, 225–229. doi:10.1097/00000433-200009000-00008

Caldicott DGE, Mahajani R, Kuhn M (2001) The anatomy of a shark attack: a case report and review of the literature. *Injury* **32**, 445–453. doi:10.1016/S0020-1383(01)00041-9

Campana SE, Joyce W, Manning MJ (2009) Bycatch and discard mortality in commercially caught blue sharks *Prionace glauca* assessed using archival satellite pop-up tags. *Marine Ecology Progress Series* **387**, 241–253. doi:10.3354/meps08109

Campana SE, Joyce W, Fowler M, Showell M (2016) Discards, hooking, and post-release mortality of porbeagle (*Lamna nasus*), shortfin mako (*Isurus oxyrinchus*), and blue shark (*Prionace glauca*) in the Canadian pelagic longline fishery. *ICES Journal of Marine Science* **73**, 520–528. doi:10.1093/icesjms/fsv234

Cardno (2015) *Shark Deterrents and Detectors: Review of Bather Protection Technologies.* Cardno (NSW/ACT) Pty Ltd, on behalf of NSW Department of Primary Industries.

Carlson JK, Ribera MM, Conrath CL, Heupel MR, Burgess GH (2010) Habitat use and movement patterns of bull sharks *Carcharhinus leucas* determined using pop-up satellite archival tags. *Journal of Fish Biology* **77**, 661–675.

Casper BM (2011) *The Ear and Hearing in Sharks, Skates, and Rays.* San Diego, Elsevier Academic Press.

Casper BM, Mann DA (2009) Field hearing measurements of the Atlantic sharpnose shark *Rhizoprionodon terraenovae*. *Journal of Fish Biology* **75**, 2768–2776. doi:10.1111/j.1095-8649.2009.02477.x

Castro JI (2011a) Family Carcharhinidae, the requiem sharks. *The Sharks of North America.* New York, Oxford University Press.

Castro JI (2011b) Family Dalatiidae, the kitefin shark, the cookiecutter sharks, and dwarf sharks. *The Sharks of North America.* Oxford, Oxford University Press.

Chapman BK, McPhee D (2016) Global shark attack hotspots: identifying underlying factors behind increased unprovoked shark bite incidence. *Ocean and Coastal Management* **133**, 72–84. doi:10.1016/j.ocecoaman.2016.09.010

Chapman CA, Harahush BK, Renshaw GMC (2011) The physiological tolerance of the grey carpet shark (*Chiloscyllium punctatum*) and the epaulette shark (*Hemiscyllium ocellatum*) to anoxic exposure at three seasonal temperatures. *Fish Physiology and Biochemistry* **37**, 387–399. doi:10.1007/s10695-010-9439-y

Charlesworth D (1976) First-aid for shark attack. *South African Nursing Journal* **43**, 24.

Chorpita BF, Barlow DH (1998) The development of anxiety: the role of control in the early environment. *Psychological Bulletin* **124**, 3–21. doi:10.1037/0033-2909.124.1.3

Cisneros-Montemayor AM, Barnes-Mauthe M, Al-Abdulrazzak D, Navarro-Holm E, Sumaila UR (2013) Global economic value of shark ecotourism: implications for conservation. *Oryx* **47**, 381–388. doi:10.1017/S0030605312001718

Clarke SC, McAllister MK, Milner-Gulland EJ, Kirkwood GP, Michielsens CGJ, Agnew DJ, Pikitch EK, Nakano H, Shivji MS (2006) Global estimates of shark catches using trade records from commercial markets. *Ecology Letters* **9**, 1115–1126. doi:10.1111/j.1461-0248.2006.00968.x

Cliff G (1991) Shark attacks on the South African coast between 1960 and 1990. *South African Journal of Science* **87**, 513–518.

Cliff G, Dudley SFJ (2011) Reducing the environmental impact of shark-control programs: a case study from KwaZulu-Natal, South Africa. *Marine and Freshwater Research* **62**, 700–709. doi:10.1071/MF10182

Clua E, Buray N, Legendre P, Mourier J, Planes S (2010) Behavioural response of sicklefin lemon sharks *Negaprion acutidens* to underwater feeding for ecotourism purposes. *Marine Ecology Progress Series* **414**, 257–266. doi:10.3354/meps08746

CNN (2013) Why *Shark Week* is so successful. *The Lead with Jake Tapper* (online). http://thelead.blogs.cnn.com/2013/08/05/why-shark-week-is-so-successful/

Collier RS (1992) Recurring attacks by white sharks on test-subjects at two Pacific sites off Mexico and California. *Environmental Biology of Fishes* **33**, 319–325. doi:10.1007/BF00005879

Collin SP (2012) The neuroecology of cartilaginous fishes: sensory strategies for survival. *Brain, Behavior and Evolution* **80**, 80–96. doi:10.1159/000339870

Compagno LJV (1984) An annotated and illustrated catalogue of shark species known to date. Part 2. Carcharhiniformes. *FAO Species Catalogue*. Rome, Food and Agriculture Organization of the UN.

Cook M, Mineka S (1990) Selective associations in the observational conditioning of fear in monkeys. *Journal of Experimental Psychology. Animal Behavior Processes* **16**, 372–389. doi:10.1037/0097-7403.16.4.372

Coppleson VM (1933) Shark attacks in Australian waters. *Medical Journal of Australia* **1**, 449–466.

Coppleson VM (1950) A review of shark attacks in Australian waters since 1919. *Medical Journal of Australia* **2**, 680–687.

Coppleson VM (1959) *Shark Attack*. Sydney, Angus and Robertson.

Cortés E (2000) Life history patterns and correlations in sharks. *Reviews in Fisheries Science* **8**, 299–344. doi:10.1080/10408340308951115

Coyne D (2016) Ballina rejects report from South African shark experts. *Echonetdaily*. http://www.echo.net.au/2016/04/ballina-rejects-report-from-south-african-shark-experts/

Cramer J (2004) Life after catch and release. *Marine Fisheries Review* **66**, 27–30.

Crossley R, Collins CM, Sutton SG, Huveneers C (2014) Public perception and understanding of shark attack mitigation measures in Australia. *Human Dimensions of Wildlife* **19**, 154–165. doi:10.1080/10871209.2014.844289

Daly R, Smale MJ, Cowley PD, Froneman PW (2014) Residency patterns and migration dynamics of adult bull sharks (*Carcharhinus leucas*) on the east coast of southern Africa. *PLoS One* **9**, e109357. doi:10.1371/journal.pone.0109357

Daly T, Peddemors V (eds) (2015) *Shark Meshing (Bather Protection) Program 2014–15: Annual Performance Report*: NSW Department of Primary Industries.

Danylchuk AJ, Suski CD, Mandelman JW, Murchie KJ, Haak CR, Brooks AML, Cooke SJ (2014) Hooking injury, physiological status and short-term mortality of juvenile lemon sharks (*Negaprion bevirostris*) following catch-and-release recreational angling. *Conservation Physiology* **2**, 1–10.

Davidson LNK, Krawchuk MA, Dulvy NK (2016) Why have global shark and ray landings declined: improved management or overfishing? *Fish and Fisheries* **17**, 438–458. doi:10.1111/faf.12119

Davies DH, Campbell GD (1962) The aetiology, clinical pathology and treatment of shark bite. *Journal of the Royal Naval Medical Service* **3**, 110–136.

Davis M (1992) The role of the amygdala in conditioned fear. In J Aggleton (ed.), *The Amygdala: Neurobiological Aspects of Emotion, Memory, and Mental Dysfunction*. New York, Wiley-Liss.

Davis M, Lee Y (1998) Fear and anxiety: possible roles of the amygdala and bed nucleus of the stria terminalis. *Cognition and Emotion* **12**, 277–305. doi:10.1080/026999398379619

Dehaene S (2008) *What are numbers really? A cerebral basis for number sense* (online). Technische Universtät Kaiserslautern, Department of Mathematics. www.mathematik.uni-kl.de/~wwwfktn/homepage/deHAENE.html

Dent F, Clarke S (2015) State of the global market for shark products. *FAO Fisheries and Aquaculture Technical Paper No. 590*. Rome, Food and Agriculture Organization of the UN.

Diaz GA, Serafy JE (2005) Longline-caught blue shark (*Prionace glauca*): factors affecting the numbers available for live release. *Fishery Bulletin* **103**, 720–724.

Dickman AJ, Dickman AJ (2010) Complexities of conflict: the importance of considering social factors for effectively resolving human–wildlife conflict. *Animal Conservation* **13**, 458–466. doi:10.1111/j.1469-1795.2010.00368.x

Diez G, Soto M, Blanco JM (2015) Biological characterization of the skin of shortfin mako shark *Isurus oxyrinchus* and preliminary study of the hydrodynamic behaviour through computational fluid dynamics. *Journal of Fish Biology* **87**, 123–137. doi:10.1111/jfb.12705

Dixon P (2002) *Final Recommendation: Current Shark Meshing Program in New South Wales Waters*. Fisheries Scientific Committee.

DoF (Department of Fisheries) (2012a) Guidelines for fishing for sharks posing an imminent threat to public safety. Perth, WA Department of Fisheries. https://www.dpc.wa.gov.au/Consultation/Documents/Appendix%203%20 Guidelines%20for%20fishing%20for%20sharks%20posing%20an%20 imminent%20threat.pdf

DoF (Department of Fisheries) (2012b) *Shark Hazard Mitigation Strategies Aimed at Improving Safety*. Perth, WA Department of Fisheries.

Dole N (2016) Ballina shark attack: NSW government committed millions to barriers not delivered. *ABC News*. http://www.abc.net.au/news/2016-09-27/ nsw-government-commits-millions-on-shark-nets-not-delivered/7880444

Domingo A, Coelho R, Cortes E, Garcia-Cortes B, Mas F, Mejuto J, Miller P, Ramos-Cartelle A, Santos MN, Yokawa K (2016) Is the tiger shark *Galeocerdo cuvier* a coastal species? Expanding its distribution range in the Atlantic Ocean using at-sea observer data. *Journal of Fish Biology* **88**, 1223–1228. doi:10.1111/jfb.12887

Dorling P (2014) Shark cull: 80% of Australians opposed, poll finds. *Sydney Morning Herald*. http://www.smh.com.au/environment/shark-cull-80-of-australians-opposed-poll-finds-20140128-31jtr.html

Douglas J, Brown P, Hunt T, Rogers M, Allen M (2010) Evaluating relative impacts of recreational fishing harvest and discard mortality on Murray cod (*Maccullochella peelii peelii*). *Fisheries Research* **106**, 18–21. doi:10.1016/j. fishres.2010.06.006

DPC (Department of the Premier and Cabinet) (2014) *Review: Western Australia Shark Hazard Mitigation Drum Line Program 2013–14*. Perth, WA Department of the Premier and Cabinet.

DPC (Department of the Premier and Cabinet) (2016) *Western Australian Shark Hazard Mitigation*. Perth, WA Department of the Premier and Cabinet.

DPI (Department of Primary Industries) (2009) Report into the NSW shark meshing (bather protection) program (public consultation document). NSW Department of Primary Industries. http://www.dpi.nsw.gov.au/data/assets/ pdffile/0008/276029/Report-into-the-NSW-Shark-Meshing-Program.pdf

Dudley SFJ (1997) A comparison of the shark control programs of New South Wales and Queensland (Australia) and KwaZulu-Natal (South Africa). *Ocean and Coastal Management* **34**, 1–27. doi:10.1016/S0964-5691(96)00061-0

Dudley SFJ, Cliff G (1993) Some effects of shark nets in the Natal nearshore environment. *Environmental Biology of Fishes* **36**, 243–255. doi:10.1007/BF00001720

Dudley SFJ, Cliff G (2010) Shark control: methods, efficacy, and ecological impact. In JC Carrier, JA Musick, MR Heithaus (eds), *Sharks and their Relatives II: Biodiversity, Adaptive Physiology, and Conservation*. Boca Raton, CRC Press.

Dudley SFJ, Gribble NA (1999) Management of shark control programs. In R Shotton (ed.), Case studies of the management of elasmobranch fishes, Part 2. *FAO Fisheries Technical Paper* **378**, 819–859.

Dudley SFJ, Simpfendorfer CA (2006) Population status of 14 shark species caught in the protective gillnets off KwaZulu-Natal beaches, South Africa, 1978–2003. *Marine and Freshwater Research* **57**, 225–240. doi:10.1071/MF05156

Duffy MA, Housley JM, Penczykowski RM, Cáceres CE, Hall SR (2011) Unhealthy herds: indirect effects of predators enhance two drivers of disease spread. *Functional Ecology* **25**, 945–953. doi:10.1111/j.1365-2435.2011.01872.x

Dulvy NK, Fowler SL, Musick JA, Cavanagh RD, Kyne PM, Harrison LR, Carlson JK, Davidson LNK, Fordham SV, Francis MP, Pollock CM, Simpfendorfer CA, Burgess GH, Carpenter KE, Compagno LJV, Ebert DA, Gibson C, Heupel MR, Livingstone SR, Sanciangco JC, Stevens JD, Valenti S, White WT (2014) Extinction risk and conservation of the world's sharks and rays. *eLife* **3**, e00590. doi:10.7554/eLife.00590

Dunham KM, Ghiurghi A, Cumbi R, Urbano F (2010) Human–wildlife conflict in Mozambique: a national perspective, with emphasis on wildlife attacks on humans. *Oryx* **44**, 185–193. doi:10.1017/S003060530999086X

DVIDS (2016) *Making a Splash in New Research with Shark Antibodies*. Defense Media Activity. https://www.dvidshub.net/news/212878/making-splash-new-research-with-shark-antibodies

Edrén SMC, Gruber SH (2005) Homing ability of young lemon sharks, *Negaprion brevirostris*. *Environmental Biology of Fishes* **72**, 267–281. doi:10.1007/s10641-004-2583-4

Eliot I, Tonts M, Eliot M, Walsh G, Collins J (2005) *Recreational Beach Users in the Perth Metropolitan Area: March 2005 in Summer 2004–2005*. Institute of Regional Development, University of Western Australia.

Epstein S (1972) The nature of anxiety with emphasis upon its relationship to expectancy. In CD Spielberger (ed.), *Anxiety: Current Trends in Theory and Research*. New York, Academic Press.

Estado de Pernambuco (1999) *Decreto Nº 21.402, de 6 de Maio de 1999* (online). Palácio do Campo das Princesas. http://legis.alepe.pe.gov.br/Paginas/Impressao/texto.aspx?nomeArquivo=DE214021999

Fanselow MS (1994) Neural organization of the defensive behavior system responsible for fear. *Psychonomic Bulletin and Review* **1**, 429–438. doi:10.3758/BF03210947

FAO FAD (Fisheries and Aquaculture Department) (2010) *The State of World Fisheries and Aquaculture*. Rome, Food and Agriculture Organization of the UN.

Federal Bureau of Prisons (2016) *Alcatraz origins* (online). https://www.bop.gov/about/history/alcatraz.jsp

Feldheim KA, Gruber SH, Ashley MV (2004) Reconstruction of parental microsatellite genotypes reveals female polyandry and philopatry in the lemon shark, *Negaprion brevirostris*. *Evolution* **58**, 2332–2342. doi:10.1111/j.0014-3820.2004.tb01607.x

Fernicola RG (2001) *Twelve Days of Terror: A Definitive Investigation of the 1916 New Jersey Shark Attacks*. Lanham, Lyons Press.

Ferreira L, Thums M, Meeuwig J, Vianna G, Stevens J, McAuley R, Meekan M (2015) Crossing latitudes: long-distance tracking of an apex predator. *PLoS One* **10**, e0116916. doi:10.1371/journal.pone.0116916

Ferretti F, Worm B, Britten GL, Heithaus MR, Lotze HK (2010) Patterns and ecosystem consequences of shark declines in the ocean. *Ecology Letters* **13**, 1055–1071.

Ferretti F, Jorgensen S, Chapple TK, De Leo G, Micheli F (2015) Reconciling predator conservation with public safety. *Frontiers in Ecology and the Environment* **13**, 412–417. doi:10.1890/150109

Fitzpatrick R, Abrantes KG, Seymour J, Barnett A (2011) Variation in depth of whitetip reef sharks: does provisioning ecotourism change their behaviour? *Coral Reefs* **30**, 569–577. doi:10.1007/s00338-011-0769-8

Fogelberg JM (1944) *Final Report on the Use of Chemical Materials as Shark Repellents*. Washington, DC, Naval Research Laboratory, Anacostia Station.

Fraser PJ, Shelmerdine RL (2002) Dogfish hair cells sense hydrostatic pressure. *Nature* **415**, 495–496. doi:10.1038/415495a

Frazzetta TH (1988) The mechanics of cutting and the form of shark teeth (Chondrichthyes, Elasmobranchii). *Zoomorphology* **108**, 93–107. doi:10.1007/BF00539785

French RP, Lyle J, Tracey S, Currie S, Semmens JM (2015) High survivorship after catch-and-release fishing suggests physiological resilience in the endothermic shortfin mako shark (*Isurus oxyrinchus*). *Conservation Physiology* **3**, cov044.

Friedrich LA, Jefferson R, Glegg G (2014) Public perceptions of sharks: gathering support for shark conservation. *Marine Policy* **47**, 1–7. doi:10.1016/j.marpol.2014.02.003

Gallagher A, Serafy J, Cooke S, Hammerschlag N (2014) Physiological stress response, reflex impairment, and survival of five sympatric shark species following experimental capture and release. *Marine Ecology Progress Series* **496**, 207–218. doi:10.3354/meps10490

Gallagher AJ, Vianna GMS, Papastamatiou YP, Macdonald C, Guttridge TL, Hammerschlag N (2015) Biological effects, conservation potential, and research priorities of shark-diving tourism. *Biological Conservation* **184**, 365–379. doi:10.1016/j.biocon.2015.02.007

Gates B (2014) The deadliest animal in the world. *gatesnotes: The Blog of Bill Gates* (online). https://www.gatesnotes.com/Health/Most-Lethal-Animal-Mosquito-Week

Gerbner G (1969) Toward 'cultural indicators': the analysis of mass mediated public message systems. *AV Communication Review* **17**, 137–148.

Gibbs L, Warren A (2014) Killing sharks: cultures and politics of encounter and the sea. *Australian Geographer* **45**, 101–107. doi:10.1080/00049182.2014.899023

Gibbs L, Warren A (2015) Transforming shark hazard policy: learning from ocean-users and shark encounter in Western Australia. *Marine Policy* **58**, 116–124. doi:10.1016/j.marpol.2015.04.014

Godin AC, Carlson JK, Burgener V (2012) The effect of circle hooks on shark catchability and at-vessel mortality rates in longline fisheries. *Bulletin of Marine Science* **88**, 469–483. doi:10.5343/bms.2011.1054

Goldman KJ, Anderson SD (1999) Space utilisation and swimming depth of white shark (*Carcharodon carcharias*) at the South Farallon Islands, central California. *Environmental Biology of Fishes* **56**, 351–364. doi:10.1023/A:1007520931105

Goldman KJ, Anderson SD, Latour RJ, Musick JA (2004) Homeothermy in adult salmon sharks, *Lamna ditropis*. *Environmental Biology of Fishes* **71**, 403–411. doi:10.1007/s10641-004-6588-9

Goodwin NM, White JA (1977) First aid for shark attack victims. *South African Medical Journal (Suid-Afrikaanse tydskrif vir geneeskunde)* **52**, 981.

Grogan ED, Lund R (2004) The origin and relationships of early Chondrichthyes. In JC Carrier, JA Musick, MR Heithaus (eds), *Biology of Sharks and their Relatives*. Boca Raton, CRC Press.

Gruber SH, Cohen JL (1978) Visual system of the elasmobranchs: state of the art 1960–1975. In ES Hodgson, RF Mathewson (eds), *Sensory Biology of Sharks, Skates and Rays*. Arlington, US Office of Naval Research.

Guidera KJ, Ogden JA, Highhouse K, Pugh L, Beatty E (1991) Shark attack. *Journal of Orthopaedic Trauma* **5**, 204–208. doi:10.1097/00005131-199105020-00015

Gururatsakul S, Gibbins D, Kearney D, Lee I (2010) Shark detection using optical image data from a mobile aerial platform. *2011 Canadian Conference on Computer and Robot Vision (CRV)*, 234–241.

Guttridge TL, Myrberg AA, Porcher IF, Sims DW, Krause J (2009) The role of learning in shark behaviour. *Fish and Fisheries* **10**, 450–469. doi:10.1111/j.1467-2979.2009.00339.x

Habegger ML, Motta PJ, Huber DR, Dean MN (2012) Feeding biomechanics and theoretical calculations of bite force in bull sharks (*Carcharhinus leucas*) during ontogeny. *Zoology (Jena, Germany)* **115**, 354–364. doi:10.1016/j.zool.2012.04.007

Haine O, Ridd P, Rowe R (2001) Range of electrosensory detection of prey by *Carcharhinus melanopterus* and *Himantura granulata. Marine and Freshwater Research* **52**, 291–296. doi:10.1071/MF00036

Handwerk B (2005) Bull shark threat: they swim where we swim. *National Geographic News* (online). http://news.nationalgeographic.com/news/2005/07/0719_050719_bullsharks.html

Harrison K, Cantor J (1999) Tales from the screen: enduring fright reactions to scary media. *Media Psychology* **1**, 97–116. doi:10.1207/s1532785xmep0102_1

Hart B (2015) What are the odds? Testing the claim that you're more likely to be struck by lightning than attacked by a shark. *Surfer* (online). http://www.surfermag.com/features/what-are-the-odds/#v0Wruk4kQ8GT1Viu.97

Hart NS, Collin SP (2015) Sharks senses and shark repellents. *Integrative Zoology* **10**, 38–64. doi:10.1111/1749-4877.12095

Hart NS, Theiss SM, Harahush BK, Collin SP (2011) Microspectrophoto metric evidence for cone monochromacy in sharks. *Naturwissenschaften* **98**, 193–201. doi:10.1007/s00114-010-0758-8

Hazin FHV, Afonso AS (2014) A green strategy for shark attack mitigation off Recife, Brazil. *Animal Conservation* **17**, 287–296. doi:10.1111/acv.12096

Hazin FHV, Burgess GH, Carvalho FC (2008) A shark attack outbreak off Recife, Pernambuco, Brazil: 1992–2006. *Bulletin of Marine Science* **82**, 199–212.

Hazin FHV, Afonso AS, De Castilho PC, Ferreira LC, Rocha BCLM (2013) Regional movements of the tiger shark, *Galeocerdo cuvier*, off northeastern Brazil: inferences regarding shark attack hazard. *Brazilian Archives of Biology and Technology* **85**, 1053–1062.

Heithaus M, Dill L, Marshall G, Buhlier B (2002) Habitat use and foraging behaviour of tiger sharks (*Galeocerdo cuvier*) in a seagrass ecosystem. *Marine Biology* **140**, 237–248. doi:10.1007/s00227-001-0711-7

Heithaus MR, Alejandro F, Wirsing AJ, Dill LM, Fourqurean JW, Derek B, Jordan T, Bejder L (2007a) State-dependent risk-taking by green sea turtles mediates top-down effects of tiger shark intimidation in a marine ecosystem. *Journal of Animal Ecology* **76**, 837–844. doi:10.1111/j.1365-2656.2007. 01260.x

Heithaus MR, Wirsing AJ, Dill LM, Heithaus LI (2007b) Long-term movements of tiger sharks satellite-tagged in Shark Bay, Western Australia. *Marine Biology* **151**, 1455–1461. doi:10.1007/s00227-006-0583-y

Heupel MR, Simpfendorfer CA, Hueter RE (2003) Running before the storm: blacktip sharks respond to falling barometric pressure associated with Tropical Storm Gabrielle. *Journal of Fish Biology* **63**, 1357–1363. doi:10. 1046/j.1095-8649.2003.00250.x

Heupel MR, Semmens JM, Hobday AJ (2006) Automated animal tracking: scales, design and deployment of listening station arrays. *Marine and Freshwater Research* **57**, 1–13. doi:10.1071/MF05091

Hiresquare (2015) How shark attacks impact tourism. https://www.hiresquare. com.au/blog/shark-attacks-tourism-impact/

Hodgson ES, Mathewson RF (1978) Electrophysiological studies of chemoreception in elasmobranchs. In ES Hodgson, RF Mathewson (eds), *Sensory Biology of Sharks, Skates, and Rays.* Arlington, US Office of Naval Research.

Hofmann MH, Northcutt RG (2012) Forebrain organization in elasmobranchs. *Brain, Behavior and Evolution* **80**, 142–151. doi:10.1159/000339874

Holden MJ (1977) Elasmobranchs. In JA Gulland (ed.), *Fish Population Dynamics.* New York, John Wiley & Sons.

Holmes B, Pepperell J, Griffiths S, Jaine F, Tibbetts I, Bennett M (2014) Tiger shark (*Galeocerdo cuvier*) movement patterns and habitat use determined by satellite tagging in eastern Australian waters. *Marine Biology* **161**, 2645–2658. doi:10.1007/s00227-014-2536-1

Hueter R, Mann DA, Maruska KP, Sisneros JA, Demski LS (2004) Sensory biology of elasmobranchs. In JC Carrier, JA Musick, MR Heithaus (eds), *Biology of Sharks and their Relatives.* Boca Raton, CRC Press.

Hueter RE, Heupel M, Heist EJ, Keeney DB (2005) Evidence of philopatry in sharks and implications for the management of shark fisheries. *Journal of Northwest Atlantic Fishery Science* **35**, 239–247. doi:10.2960/J.v35.m493

Hugh R, Gilardi GL (1980) *Pseudomonas.* In EH Lennette, A Balows, WJ Hausler, JP Truant (eds), *Manual of Clinical Microbiology.* 3rd edn. Washington, DC, American Society for Microbiology.

Huveneers C, Rogers PJ, Semmens JM, Beckmann C, Kock AA, Page B, Goldsworthy SD (2013) Effects of an electric field on white sharks: *in situ* testing of an electric deterrent. *PLoS One* **8**, e62730. doi:10.1371/journal. pone.0062730

Interaminense JA, Nascimento DCO, Ventura RF, Batista JEC, Souza MMC, Hazin FHV, Pontes NT, Lima JV (2010) Recovery and screening for antibiotic susceptibility of potential bacterial pathogens from the oral cavity of shark species involved in attacks on humans in Recife, Brazil. *Journal of Medical Microbiology* **59**, 941–947. doi:10.1099/jmm.0.020453-0

ISAF (2017) International shark attack file. Florida Museum, University of Florida. https://www.flmnh.ufl.edu/fish/isaf/home/

Işcan MY, McCabe BQ (1995) Analysis of human remains recovered from a shark. *Forensic Science International* 72, 15–23. doi:10.1016/0379-0738 (94)01643-J

ITU (2015) ITU releases 2015 ICT figures: statistics confirm ICT revolution of the past 15 years. Geneva. http://www.itu.int/net/pressoffice/press_releases/2015/17.aspx#.V7pZFKMkrIW

IUCN (2016) The IUCN red list of threatened species. http://www.iucnredlist.org/

Jewell OJD, Johnson RL, Gennari E, Bester MN (2013) Fine scale movements and activity areas of white sharks (*Carcharodon carcharias*) in Mossel Bay, South Africa. *Environmental Biology of Fishes* **96**, 881–894. doi:10.1007/s10641-012-0084-4

Johnson R, Kock A (2006) South Africa's white shark cage-diving industry: is there cause for concern? In DC Nel, T Peshak (eds), *Finding a Balance: White Shark Conservation and Recreational Safety in the Inshore Waters of Cape Town, South Africa*. Cape Town, WWF South Africa Report Series.

Jordan LK (2008) Comparative morphology of stingray lateral line canal and electrosensory systems. *Journal of Morphology* **269**, 1325–1339. doi:10.1002/jmor.10660

Jordan LK, Mandelman JW, Kajiura SM (2011) Behavioral responses to weak electric fields and a lanthanide metal in two shark species. *Journal of Experimental Marine Biology and Ecology* **409**, 345–350. doi:10.1016/j.jembe.2011.09.016

Jorgensen SJ, Reeb CA, Chapple TK, Anderson S, Perle C, Van Sommeran SR, Fritz-Cope C, Brown AC, Klimley AP, Block BA (2010) Philopatry and migration of Pacific white sharks. *Proceedings. Biological Sciences* **277**, 679–688. doi:10.1098/rspb.2009.1155

Kajiura SM, Holland KN (2002) Electroreception in juvenile scalloped hammerhead and sandbar sharks. *Journal of Experimental Biology* **205**, 3609–3621.

Kajiura SM, Tellman SL (2016) Quantification of massive seasonal aggregations of blacktip sharks (*Carcharhinus limbatus*) in southeast Florida. *PLoS One* **11**, e0150911. doi:10.1371/journal.pone.0150911

Kalmijn AJ (1982) Electric and magnetic field detection in elasmobranch fishes. *Science* **218**, 916–918. doi:10.1126/science.7134985

Kapp BS, Whalen P, Supple WF, Pascoe JP (1992) Amygdaloid contributions to conditioned arousal and sensory information processing. In JP Aggleton (ed.), *The Amygdala: Neurobiological Aspects of Emotion, Memory, and Mental Dysfunction*. New York, Wiley-Liss.

Kempster RM, Egeberg CA, Hart NS, Ryan L, Chapuis L, Kerr CC, Schmidt C, Huveneers C, Gennari E, Yopak KE, Meeuwig JJ, Collin SP (2016) How close is too close? The effect of a non-lethal electric shark deterrent on white shark behaviour. *PLoS One* **11**, e0157717. doi:10.1371/journal. pone.0157717

Kimber JA, Sims DW, Bellamy PH, Gill AB (2014) Elasmobranch cognitive ability: using electroreceptive foraging behaviour to demonstrate learning, habituation and memory in a benthic shark. *Animal Cognition* **17**, 55–65. doi:10.1007/s10071-013-0637-8

Klimley AP, Myrberg AA Jr (1979) Acoustic stimuli underlying withdrawal from a sound source by adult lemon sharks, *Negaprion brevirostris* (Poey). *Bulletin of Marine Science* **29**, 447–458.

Kock A, Johnson R (2006) White shark abundance: not a causative factor in numbers of shark bite incidents. In DC Nel, TP Peschak (eds), *Finding a Balance: White Shark Conservation and Recreational Safety in the Inshore Waters of Cape Town, South Africa*. Cape Town, WWF South Africa Report Series.

Kock A, O'Riain MJ, Mauff K, Meyer M, Kotze D, Griffiths C (2013) Residency, habitat use and sexual segregation of white sharks, *Carcharodon carcharias* in False Bay, South Africa. *PLoS One* **8**, e55048. doi:10.1371/journal.pone.0055048

Krajacic S (2016) Shark attack survivor Brett Connellan positive about latest deterrence technology. *ABC Illawarra* (online). http://www.abc.net.au/local/stories/2016/07/29/4509795.htm

Lack M, Sant G (2009) *Trends in Global Shark Catch and Recent Developments in Management*. Cambridge, TRAFFIC International.

Lack M, Sant G (2011) The future of sharks: a review of action and inaction. *TRAFFIC report*. Cambridge: TRAFFIC International and Pew Environment Group.

Lai M-C, Yang S-N (2011) Perinatal hypoxic-ischemic encephalopathy. *Journal of Biomedicine and Biotechnology* **2011**, 609813. doi:10.1155/2011/ 609813

Langley RL (2005) Animal-related fatalities in the United States: an update. *Wilderness and Environmental Medicine* **16**, 67–74. doi:10.1580/1080-6032(2005)16[67:AFITUS]2.0.CO;2

Larkin PA (1979) Predator–prey relations in fishes: an overview of the theory. In H Clepper (ed.), *Predator–Prey Systems in Fisheries Management*. Washington, DC, Sport Fishing Institute.

Laroche RK, Kock AA, Dill LM, Oosthuizen WH (2007) Effects of provisioning ecotourism activity on the behaviour of white sharks

Carcharodon carcharias. Marine Ecology Progress Series **338**, 199–209. doi:10.3354/meps338199

Lea JSE, Humphries NE, Clarke CR, Sims DW (2015) To Madagascar and back: long-distance, return migration across open ocean by a pregnant female bull shark *Carcharhinus leucas. Journal of Fish Biology* **87**, 1313–1321. doi:10.1111/jfb.12805

Lentz AK, Burgess GH, Perrin K, Brown JA, Mozingo DW, Lottenberg L (2010) Mortality and management of 96 shark attacks and development of a shark bite severity scoring system. *American Surgeon* **76**, 101–106.

Levine M (1996) Unprovoked attacks by white sharks off the South African coast. In AP Klimely, DG Ainley (eds), *Great White Sharks: The Biology of Carcharodon carcharias.* San Diego, Academic Press.

Levine M, Collier RS, Ritter E, Fouda M, Canabal V (2014) Shark cognition and a human-mediated driver of a spate of shark attacks. *Open Journal of Animal Sciences* **4**, 263–269. doi:10.4236/ojas.2014.45033

Lewison R, Oliver W (2008) *Hippopotamus amphibius.* IUCN SSC Hippo Specialist Subgroup. IUCN Red List of Threatened Species 2008. http://dx.doi.org/10.2305/IUCN.UK.2008.RLTS.T10103A3163790.en

Li CH, Corrigan S, Yang L, Straube N, Harris M, Hofreiter M, White WT, Naylor GJP (2015) DNA capture reveals transoceanic gene flow in endangered river sharks. *Proceedings of the National Academy of Sciences of the United States of America* **112**, 13302–13307. doi:10.1073/pnas.1508735112

Lindberg DR, Pyenson ND (2006) Evolutionarpatterns in Cetacea: fishing up prey size through deep time. In JA Estes, DP DeMaster, DF Doak, TM Williams, RL Brownell (eds), *Whales, Whaling, and Ocean Ecosystems.* Los Angeles, University of California Press.

Löe J, Röskaft E (2004) Large carnivores and human safety: a review. *Ambio* **33**, 283–288. doi:10.1579/0044-7447-33.6.283

Loewenstein G, Mather J (1990) Dynamic processes in risk perception. *Journal of Risk and Uncertainty* **3**, 155–175. doi:10.1007/BF00056370

Loewenstein GF, Weber EU, Hsee CK, Welch ES (2001) Risk as feelings. *Psychological Bulletin* **127**, 267–286. doi:10.1037/0033-2909.127.2.267

Lowe CG, Goldman KJ (2001) Thermal and bioenergetics of elasmobranchs: bridging the gap. *Environmental Biology of Fishes* **60**, 251–266. doi:10.1023/A:1007650502269

Lubow RE (1998) Latent inhibition and behavior pathology: prophylactic and other possible effects of stimulus preexposure. In W O'Donohue (ed.), *Learning and Behavior Therapy.* Needham Heights, Allyn & Bacon.

Maisey JG (2001) Remarks on the inner ear of elasmobranchs and its interpretation from skeletal labyrinth morphology. *Journal of Morphology* **250**, 236–264. doi:10.1002/jmor.1068

Maljković A, Côté IM (2011) Effects of tourism-related provisioning on the trophic signatures and movement patterns of an apex predator, the Caribbean reef shark. *Biological Conservation* **144**, 859–865. doi:10.1016/j. biocon.2010.11.019

Marie C-V, Rallu J-L (2012) *Demographic and Migration Trends in the Outermost Regions: Impacts on Territorial, Social and Territorial Cohesion? Executive Summary – Reunion Island*. European Commission, Directorate General for Regional Policy.

Marketforce (2013) Western Australian beaches: promoting safe beach use. Perth. http://www.parliament.wa.gov.au/publications/tabledpapers.nsf/display paper/3912002ceb1438c6206bfc7348257d5e000ec722/$file/tp-2002.pdf

Marra NJ, Richards VP, Early A, Bogdanowicz SM, Pavinski Bitar PD, Stanhope MJ, Shivji MS (2017) Comparative transcriptomics of elasmobranchs and teleosts highlight important processes in adaptive immunity and regional endothermy. *BMC Genomics* **18**, 87. doi:10.1186/ s12864-016-3411-x

Marshall H, Field L, Afiadata A, Sepulveda C, Skomal G, Bernal D (2012) Hematological indicators of stress in longline-captured sharks. *Comparative Biochemistry and Physiology. Part A* **162**, 121–129.

Martin JA (2016) Seeing jaws: the role of shark science in ocean conservation. *Historical Studies in the Natural Sciences* **46**, 67–100. doi:10. 1525/ hsns.2016.46.1.67

Matich P, Heithaus MR (2015) Individual variation in ontogenetic niche shifts in habitat use and movement patterns of a large estuarine predator (*Carcharhinus leucas*). *Oecologia* **178**, 347–359. doi:10.1007/s00442-015-3253-2

McAuley R, Bruce B, Keaya I, Mountforda S, Pinnella T (2016) *Evaluation of Passive Acoustic Telemetry Approaches for Monitoring and Mitigating Shark Hazards off the Coast of Western Australia*. Perth, WA Department of Fisheries.

McCagh C, Sneddon J, Blache D (2015) Killing sharks: the media's role in public and political response to fatal human–shark interactions. *Marine Policy* **62**, 271–278. doi:10.1016/j.marpol.2015.09.016

McComb DM, Tricas TC, Kajiura SM (2009) Enhanced visual fields in hammerhead sharks. *Journal of Experimental Biology* **212**, 4010–4018. doi:10.1242/jeb.032615

McCosker JE (1985) White shark attack behaviour: observations and speculations about predator and prey tactics. *Bulletin of the Southern California Academy of Sciences* **9**, 123–135.

McCutcheon SM, Kajiura SM (2013) Electrochemical properties of lanthanide metals in relation to their application as shark repellents. *Fisheries Research* **147**, 47–54. doi:10.1016/j.fishres.2013.04.014

McFadden EB, Johnson CS (1978) Color and reflectivity of sea-survival equipment as related to shark attack. In EB McFadden (ed.), *Flotation and Survival Equipment Studies.* Oklahoma City, FAA Civil Aeromedical Institute.

McPhee D (2012) *Likely Effectiveness of Netting or Other Capture Programs as a Shark Hazard Mitigation Strategy in Western Australia.* Perth.

McPhee D (2014) Unprovoked shark bites: are they becoming more prevalent? *Coastal Management* 42, 478–492. doi:10.1080/08920753.2014.942046

Meadows E (1999) Aboriginal history of the Waverley area: a discussion paper. Reference Department, Waverly Library. http://www.waverley.nsw.gov. au/__data/assets/pdf_file/0010/8659/Aboriginal_History_of_the_ Waverley_Area.pdf

Meyer W, Seegers U (2012) Basics of skin structure and function in elasmobranchs: a review. *Journal of Fish Biology* 80, 1940–1967. doi:10.1111/ J.1095-8649.2011.03207.x

Meyer C, Clark T, Papastamatiou Y, Whitney N, Holland K (2009a) Long-term movement patterns of tiger sharks *Galeocerdo cuvier* in Hawaii. *Marine Ecology Progress Series* 381, 223–235. doi:10.3354/meps07951

Meyer CG, Dale JJ, Papastamatiou YP, Whitney NM, Holland KN (2009b) Seasonal cycles and long-term trends in abundance and species composition of sharks associated with cage-diving ecotourism activities in Hawaii. *Environmental Conservation* 36, 104–111. doi:10.1017/S0376892909990038

Meyer CG, Papastamatiou YP, Holland KN (2010) A multiple instrument approach to quantifying the movement patterns and habitat use of tiger (*Galeocerdo cuvier*) and Galapagos sharks (*Carcharhinus galapagensis*) at French Frigate Shoals, Hawaii. *Marine Biology* 157, 1857–1868. doi:10. 1007/s00227-010-1457-x

Miller DJ, Collier RS (1980) Shark attacks in California and Oregon, 1926–1979. *California Fish and Game* 67, 76–104.

Mineka S (1992) Evolutionary memories, emotional processing, and the emotional disorders. *Psychology of Learning and Motivation* 28, 161–206. doi:10.1016/S0079-7421(08)60490-9

Mineka S, Zinbarg R (2006) A contemporary learning theory perspective on the etiology of anxiety disorders: it's not what you thought it was. *American Psychologist* 61, 10–26. doi:10.1037/0003-066X.61.1.10

Mineka S, Cook M, Miller S (1984) Fear conditioned with escapable and inescapable shock: the effects of a feedback stimulus. *Journal of Experimental Psychology. Animal Behavior Processes* 10, 307–323. doi:10. 1037/0097-7403. 10.3.307

Mitchell H (2015) *Submission #70 to the Parliamentary Inquiry into Management of Sharks in New South Wales Waters.* Submission on behalf of the Australian Aerial Patrol.

Molina JM, Cooke SJ (2012) Trends in shark bycatch research: current status and research needs. *Reviews in Fish Biology and Fisheries* 22, 719–737. doi:10.1007/s11160-012-9269-3

Morgan M, Shanahan J (2010) The state of cultivation. *Journal of Broadcasting and Electronic Media* 54, 337–355. doi:10.1080/08838151003735018

Motta PJ (2004) Prey capture behaviour and feeding mechanics of elasmobranchs. In JC Carrier, JA Musick, MR Heithaus (eds), *Biology of Sharks and their Relatives*. Boca Raton, CRC Press.

Murtugudde R, McCreary JP Jr, Busalacchi AJ (2000) Oceanic processes associated with anomalous events in the Indian Ocean with relevance to 1997–1998. *Journal of Geophysical Research* 105, 3295–3306. doi:10.1029/1999JC900294

Muter BA, Gore ML, Gledhill KS, Lamont C, Huveneers C (2013) Australian and US news media portrayal of sharks and their conservation. *Conservation Biology* 27, 187–196. doi:10.1111/j.1523-1739.2012.01952.x

Myers RA, Baum JK, Shepherd TD, Powers SP, Peterson CH (2007) Cascading effects of the loss of apex predatory sharks from a coastal ocean. *Science* 315, 1846–1850. doi:10.1126/science.1138657

Myrberg AA Jr, Banner A, Richard JD (1969) Shark attraction using a video-acoustic system. *Marine Biology* 2, 264–276. doi:10.1007/BF00351149

Myrberg AA Jr, Gordon CR, Klimley AP (1978) Rapid withdrawal from a sound source by open-ocean sharks. *Journal of the Acoustical Society of America* 64, 1289–1297. doi:10.1121/1.382114

Myrick JG, Evans SD (2014) Do PSAs take a bite out of shark week? The effects of juxtaposing environmental messages with violent images of shark attacks. *Science Communication* 36, 544–569. doi:10.1177/1075547014547159

Neff C (2012) Australian beach safety and the politics of shark attacks. *Coastal Management* 40, 88–106. doi:10.1080/08920753.2011.639867

Neff C (2014) The politics of shark attacks. *Save Our Seas Foundation Magazine*. Geneva, Save Our Seas Foundation.

Neff C (2015) The *Jaws* effect: how movie narratives are used to influence policy responses to shark bites in Western Australia. *Australian Journal of Political Science* 50, 114–127. doi:10.1080/10361146.2014.989385

Neff C, Hueter R (2013) Science, policy, and the public discourse of shark 'attack': a proposal for reclassifying human–shark interactions. *Journal of Environmental Studies and Sciences* 3, 65–73. doi:10.1007/s13412-013-0107-2

Nel DC, Peschak TP (eds) (2006) *Finding a Balance: White Shark Conservation and Recreational Safety in the Inshore Waters of Cape Town, South Africa*. Cape Town, WWF Sanlam Marine Programme, WWF South Africa.

Nielsen J, Hedeholm RB, Heinemeier J, Bushnell PG, Christiansen JS, Olsen J, Ramsey CB, Brill RW, Simon M, Steffensen KF, Steffensen JF (2016) Eye lens radiocarbon reveals centuries of longevity in the Greenland shark (*Somniosus microcephalus*). *Science* **353**, 702–704. doi:10. 1126/science. aaf1703

Nosal AP, Keenan EA, Hastings PA, Gneezy A (2016) The effect of background music in shark documentaries on viewers' perceptions of sharks. *PLoS One* **11**, e0159279. doi:10.1371/journal.pone.0159279

O'Connell CP, de Jonge VN (2014) Integrating the findings from this special issue and suggestions for future conservation efforts: a brief synopsis. *Ocean and Coastal Management* **97**, 58–60. doi:10.1016/j.ocecoaman.2014.05.022

O'Connell CP, Gruber SH, Abel DC, Stroud EM, Rice PH (2011) The responses of juvenile lemon sharks, *Negaprion brevirostris*, to a magnetic barrier. *Ocean and Coastal Management* **54**, 225–230. doi:10.1016/j.ocecoaman.2010.11.006

O'Connell CP, Andreotti S, Rutzen M, Meyer M, Matthee CA, He PG (2014) Effects of the Sharksafe barrier on white shark (*Carcharodon carcharias*) behavior and its implications for future conservation technologies. *Journal of Experimental Marine Biology and Ecology* **460**, 37–46. doi:10.1016/j.jembe.2014.06.004

O'Gower AK (1995) Speculations on a spatial memory for the Port-Jackson shark (*Heterodontus portusjacksoni*) (Meyer) (Heterodontidae). *Marine and Freshwater Research* **46**, 861–871. doi:10.1071/MF9950861

Öhman A, Mineka S (2001) Fears, phobias, and preparedness: toward an evolved module of fear and fear learning. *Psychological Review* **108**, 483–522. doi:10.1037/0033-295X.108.3.483

Öhman A, Dimberg U, Öst L-G (1985) Animal and social phobias: biological constraints on learned fear responses. In S Reiss, RR Bootzin (eds), *Theoretical Issues in Behavior Therapy*. New York, Academic Press.

Olsson A, Phelps EA (2007) Social learning of fear. *Nature Neuroscience* **10**, 1095–1102. doi:10.1038/nn1968

Orams MB (2002) Feeding wildlife as a tourism attraction: a review of issues and impacts. *Tourism Management* **23**, 281–293. doi:10.1016/S0261-5177(01)00080-2

Ozyilmaz A, Oksuz A (2015) Determination of the biochemical properties of liver oil from selected cartilaginous fish living in the northeastern Mediterranean. *Journal of Animal and Plant Sciences* **25**, 160–167.

Paddenburg T (2016) WA shark study questions affect of tagging on animals' feeding ability. *Perth Now*. http://www.perthnow.com.au/news/western-australia/wa-shark-study-questions-affect-of-tagging-on-animals-feeding-ability/news-story/535c6d51fa06b293d0addb17cd7d9c27

Papastamatiou YP, Cartamil DP, Lowe CG, Meyer CG, Wetherbee BM, Holland KN (2011) Scales of orientation, directed walks and movement path structure in sharks. *Journal of Animal Ecology* **80**, 864–874. doi:10.1111/j.1365-2656.2011.01815.x

Papastamatiou YP, Meyer CG, Carvalho F, Dale JJ, Hutchinson MR, Holland KN (2013) Telemetry and random walk models reveal complex patterns of partial migration in a large marine predator. *Ecology* **94**, 2595–2606. doi:10.1890/12-2014.1

Papson S (1992) 'Cross the fin line of terror' *Shark Week* on the Discovery Channel. *Journal of American Culture* **15**, 67. doi:10.1111/j.1542-734X. 1992.1504_67.x

Parrish FA, Goto RS (1997) Patterns of insular shark dynamics based on fishery bycatch and lifeguard surveillance at Oahu, Hawaii, 1983–1992. *Bulletin of Marine Science* **61**, 763–777.

Parsons MJG, Parnum IM, Allen K, McCauley RD, Erbe C (2014) Detection of sharks with the Gemini imaging sonar. *Acoustics Australia* **42**, 185–189.

Paterson R (1990) Effects of long-term anti-shark measures on target and non-target species in Queensland, Australia. *Biological Conservation* **52**, 147–159. doi:10.1016/0006-3207(90)90123-7

Pavia AT, Bryan JA, Maher KL, Hester TR Jr, Farmer JJ (1989) *Vibrio carchariae* infection after a shark bite. *Annals of Internal Medicine* **111**, 85–86. doi:10. 7326/0003-4819-111-1-85

Peace A (2015) Shark attack! A cultural approach. *Anthropology Today* **31**, 3–7. doi:10.1111/1467-8322.12197

Perni M, Galvagnion C, Maltsev A, Meisl G, Müller MBD, Challa PK, Kirkegaard JB, Flagmeier P, Cohen SIA, Cascella R, Chen SW, Limboker R, Sormanni P, Heller GT, Aprile FA, Cremades N, Cecchi C, Chiti F, Nollen EAA, Knowles TPJ, Vendruscolo M, Bax A, Zasloff M, Dobson CM (2017) A natural product inhibits the initiation of α-synuclein aggregation and suppresses its toxicity. *Proceedings of the National Academy of Sciences* **114**, E1009–E1017. doi:10.1073/pnas.1610586114

Peschak TP (2006) Sharks and shark bite in the media. In DC Nel, TP Peschak (eds), *Finding a Balance: White Shark Conservation and Recreational Safety in the Inshore Waters of Cape Town, South Africa; Proceedings of a Specialist Workshop*. Cape Town, WWF South Africa Report Series, 2006/Marine/001.

Phoenix Australia (2017) *Recovery and Research* (online). University of Melbourne, Phoenix Australia, Centre for Post-traumatic Mental Health. http://phoenixaustralia.org/

Pillans RD, Good JP, Anderson WG, Hazon N, Franklin CE (2005) Freshwater to seawater acclimation of juvenile bull sharks (*Carcharhinus leucas*): plasma osmolytes and Na+/K+-ATPase activity in gill, rectal gland, kidney and intestine. *Journal of Comparative Physiology. Part B* **175**, 37–44.

Pimiento C, Balk MA (2015) Body-size trends of the extinct giant shark *Carcharocles megalodon*: a deep-time perspective on marine apex predators. *Paleobiology* **41**, 479–490. doi:10.1017/pab.2015.16

Powers SP, Fodrie FJ, Scyphers SB, Drymon JM, Shipp RL, Stunz GW (2013) Gulf-wide decreases in the size of large coastal sharks documented by generations of fishermen. *Marine and Coastal Fisheries* **5**, 93–102. doi:10.10 80/19425120.2013.786001

Rachman S (1977) The conditioning theory of fear acquisition: a critical examination. *Behaviour Research and Therapy* **15**, 375–387. doi:10.1016/0005-7967(77)90041-9

Reid DD, Krogh M (1992) Assessment of catches from protective shark meshing of New South Wales between 1950 and 1990. *Australian Journal of Marine and Freshwater Research* **43**, 283–296. doi:10.1071/MF9920283

Reid DD, Robbins WD, Peddemors VM (2011) Decadal trends in shark catches and effort from the New South Wales, Australia, shark meshing program 1950–2010. *Marine and Freshwater Research* **62**, 676–693. doi:10.1071/MF10162

Renshaw GMC, Kutek AK, Grant GD, Anoopkumar-Dukie S (2012) Forecasting elasmobranch survival following exposure to severe stressors. *Comparative Biochemistry and Physiology. A. Comparative Physiology* **162**, 101–112. doi:10.1016/j.cbpa.2011.08.001

Ricci JA, Vargas CR, Singhal D, Lee BT (2016) Shark attack-related injuries: epidemiology and implications for plastic surgeons. *Journal of Plastic, Reconstructive and Aesthetic Surgery* **69**, 108–114. doi:10.1016/j.bjps.2015.08.029

Ritter E, Amin R (2012) Effect of human body position on the swimming behavior of bull sharks, *Carcharhinus leucas. Society and Animals* **20**, 225–235. doi:10.1163/15685306-12341235

Ritter EK, Amin R (2014) Are Caribbean reef sharks, *Carcharhinus perezi,* able to perceive human body orientation? *Animal Cognition* **17**, 745–753. doi:10.1007/s10071-013-0706-z

Robbins J (2016) See Florida teen bitten while wearing special shark-repellent band he got for Christmas. *International Business Times*, 30 December. http://www.ibtimes.co.uk/see-florida-teen-bitten-while-wearing-special-shark-repellent-band-he-got-christmas-1598699

Robbins WD, Peddemors VM, Kennelly SJ, Ives MC (2014) Experimental evaluation of shark detection rates by aerial observers. *PLoS One* **9**, e83456. doi:10.1371/journal.pone.0083456

Roland FP (1970) Leg gangrene and endotoxin shock due to *Vibrio parahae-molyticus*: an infection acquired in New England coastal waters. *New England Journal of Medicine* **282**, 1306. doi:10.1056/NEJM197006042822306

Rosen JB, Schulkin J (1998) From normal fear to pathological anxiety. *Psychological Review* **105**, 325–350. doi:10.1037/0033-295X.105.2.325

Rottenstreich Y, Hsee CK (2001) Money, kisses, and electric shocks: on the affective psychology of risk. *Psychological Science* **12**, 185–190. doi:10.1111/1467-9280.00334

Royle JA, Isaacs D, Eagles G, Cass D, Gilroy N, Chen S, Malouf D, Griffiths C (1997) Infections after shark attacks in Australia. *Journal of Pediatric Infectious Diseases* **16**, 531–532.

Rtshiladze MA, Andersen SP, Nguyen DQ, Grabs A, Ho K (2011) The 2009 Sydney shark attacks: case series and literature review. *ANZ Journal of Surgery* **81**, 345–351. doi:10.1111/j.1445-2197.2010.05640.x

Ryan C (1988) Saltwater crocodiles as tourist attractions. *Journal of Sustainable Tourism* **6**, 315–327.

Salamastrakis A (2017) World first trial of shark-inspired drug (online). Melbourne, La Trobe University. http://www.latrobe.edu.au/news/articles/2017/release/world-first-trial-of-shark-inspired-drug

Schmitz OJ, Grabowski JH, Peckarsky BL, Preisser EL, Trussell GC, Vonesh JR (2008) From individuals to ecosystem function: toward an integration of evolutionary and ecosystem ecology. *Ecology* **89**, 2436–2445. doi:10.1890/07-1030.1

Shark Research Institute (2016) Global shark attack file. http://www.sharkattackfile.net/incidentlog.htm

Sibert J, Hampton J, Kleiber P, Maunder M (2006) Biomass, size, and trophic status of top predators in the Pacific Ocean. *Science* **314**, 1773–1776. doi:10.1126/science.1135347

Simpfendorfer CA, Heupel MR, White WT, Dulvy NK (2011) The importance of research and public opinion to conservation management of sharks and rays: a synthesis. *Marine and Freshwater Research* **62**, 518–527. doi:10.1071/MF11086

Sisneros JA, Nelson DR (2001) Surfactants as chemical shark repellents: past, present, and future. *Environmental Biology of Fishes* **60**, 117–130. doi:10.1023/A:1007612002903

Skomal GB, Mandelman JW (2012) The physiological response to anthropogenic stressors in marine elasmobranch fishes: a review with a focus on the secondary response. *Comparative Biochemistry and Physiology. Part A* **162**, 146–155.

Slovic P, Fischhoff B, Lichtenstein S (1978) Facts versus fears: understanding perceived risk. In D Kahneman, P Slovic, A Tversky (eds), *Judgement Under Uncertainty: Heuristics and Biases*. Cambridge, Cambridge University Press.

Slovic P, Finucane ML, Peters E, MacGregor DG (2004) Risk as analysis and risk as feelings: some thoughts about affect, reason, risk, and rationality. *Risk Analysis* **24**, 311–322. doi:10.1111/j.0272-4332.2004.00433.x

Smith AM, Maguire-Nguyen KK, Rando TA, Zasloff MA, Strange KB, Yin VP (2017) The protein tyrosine phosphatase 1B inhibitor MSI-1436 stimulates regeneration of heart and multiple other tissues. *Regenerative Medicine*, **2**, article 4. doi:10.1038/s41536-017-0008-1

Smith LJ (1991) The effectiveness of sodium lauryl sulphate as a shark repellent in a laboratory test situation. *Journal of Fish Biology* **38**, 105–113. doi:10.1111/j.1095-8649.1991.tb03096.x

Smoothey AF, Gray CA, Kennelly SJ, Masens OJ, Peddemors VM, Robinson WA (2016) Patterns of occurrence of sharks in Sydney Harbour, a large urbanised estuary. *PLoS One* **11**, e0146911. doi:10.1371/journal.pone.0146911

Southall EJ, Sims DW (2003) Shark skin: a function in feeding. *Proceedings. Biological Sciences* **270**, S47–S49. doi:10.1098/rsbl.2003.0006

Springer S (1955) Laboratory experiments with shark repellents. *Proceedings of the Seventh Annual Gulf and Caribbean Fisheries Institute.* Coral Gables, FLA, 159–164.

State of Hawaii (2010) Bill text: HI HB2664: 2010: regular session: introduced (online). https://legiscan.com/HI/comments/HB2664/2010

Stillwell CE, Kohler NE (1982) Food, feeding habits, and estimates of daily ration of the shortfin mako (*Isurus oxyrinchus*) in the northwest Atlantic. *Canadian Journal of Fisheries and Aquatic Sciences* **39**, 407–414. doi:10.1139/f82-058

Stone D (1989) Causal stories and the formation of policy agendas. *Political Science Quarterly* **104**, 281–300. doi:10.2307/2151585

Stringer J, Richardson J (1980) Managing the political agenda: problem definition and policymaking in Britain. *Parliamentary Affairs* **33**, 23–39. doi:10.1093/oxfordjournals.pa.a051831

Stroud EM, O'Connell CP, Rice PH, Snow NH, Barnes BB, Elshaer MR, Hanson JE (2014) Chemical shark repellent: myth or fact? The effect of a shark necromone on shark feeding behavior. *Ocean and Coastal Management* **97**, 50–57. doi:10.1016/j.ocecoaman.2013.01.006

Sudo S, Tsuyuki K, Ito Y, Ikohagi T (2002) A study on the surface shape of fish scales. *JSME International Journal. Series C, Mechanical Systems, Machine Elements and Manufacturing* **45**, 1100–1105. doi:10.1299/jsmec.45.1100

Sunstein CR, Zeckhauser R (2008) *Overreaction to Fearsome Risks.* John M. Olin Program in Law and Economics Working Paper No. 446. University of Chicago Law School.

Tester AL, Nelson GJ, Daniels CI (1968) *Test of NUWC Shark Attack Deterrent Device.* Pasadena, CA, Naval Undersea Warfare Center.

Tricas TC (2001) The neuroecology of the elasmobranch electrosensory world: why peripheral morphology shapes behavior. *Environmental Biology of Fishes* **60**, 77–92. doi:10.1023/A:1007684404669

Unger NR, Ritter E, Borrego R, Goodman J, Osiyemi OO (2014) Antibiotic susceptibilities of bacteria isolated within the oral flora of Florida blacktip sharks: guidance for empiric antibiotic therapy. *PLoS One* **9**, e104577. doi:10.1371/journal.pone.0104577

Walsh CJ, Luer CA, Bodine AB, Smith CA, Cox HL, Noyes DR, Gasparetto M (2006) Elasmobranch immune cells as a source of novel tumor cell inhibitors: implications for public health. *Integrative and Comparative Biology* **46**, 1072–1081. doi:10.1093/icb/icl041

Weibel RE (2005) Safety considerations for operation of unmanned aerial vehicles in the national airspace system. MSc thesis, Massachusetts Institute of Technology.

Weigmann S (2016) Annotated checklist of the living sharks, batoids and chimaeras (Chondrichthyes) of the world, with a focus on biogeographical diversity. *Journal of Fish Biology* **88**, 837–1037. doi:10.1111/jfb.12874

Werry JM, Planes S, Berumen ML, Lee KA, Braun CD, Clua E (2014) Reef-fidelity and migration of tiger sharks, *Galeocerdo cuvier*, across the Coral Sea. *PLoS One* **9**, e83249. doi:10.1371/journal.pone.0083249

West JG (2011) Changing patterns of shark attacks in Australian waters. *Marine and Freshwater Research* **62**, 744–754. doi:10.1071/MF10181

Wetherbee B, Lowe C, Crow G (1994) A review of shark control in Hawaii with recommendations for future research. *Pacific Science* **48**, 95–115.

White JA (1975) Shark attack in Natal. *Injury* **6**, 187–194. doi:10.1016/0020-1383(75)90102-3

WHO (World Health Organization) (2016) *Facts about Injuries: Drowning.* Geneva, Department of Injuries and Violence Prevention, World Health Organization.

Wilson M (2010) Adaptive responses to risk and the irrationally emotional public. *Saint Louis University Law Journal* **54**, 1297–1312.

Wilson SM, Raby GD, Burnett NJ, Hinch SG, Cooke SJ (2014) Looking beyond the mortality of bycatch: sublethal effects of incidental capture on marine animals. *Biological Conservation* **171**, 61–72. doi:10.1016/j.biocon.2014.01.020

Wirsing AJ, Heithaus MR, Dill LM (2007) Can you dig it? Use of excavation, a risky foraging tactic, by dugongs is sensitive to predation danger. *Animal Behaviour* **74**, 1085–1091. doi:10.1016/j.anbehav.2007.02.009

Woods B (2000) Beauty and the beast: preferences for animals in Australia. *Journal of Tourism Studies* **11**, 25–35.

Woolgar JD, Cliff G, Nair R, Hafez H, Robbs JV (2001) Shark attack: review of 86 consecutive cases. *Journal of Trauma* **50**, 887. doi:10.1097/00005373-200105000-00019

Worm B, Davis B, Kettemer L, Ward-Paige CA, Chapman D, Heithaus MR, Kessel ST, Gruber SH (2013) Global catches, exploitation rates, and rebuilding options for sharks. *Marine Policy* **40**, 194–204. doi:10.1016/j.marpol.2012.12.034

Yopak KE (2012) Neuroecology of cartilaginous fishes: the functional implications of brain scaling. *Journal of Fish Biology* **80**, 1968–2023. doi:10.1111/j.1095-8649.2012.03254.x

Young E (2017) High hazard. *WAtoday.com.au*, 10 January. http://www.watoday.com.au/interactive/2017/sharks/high-hazard/

Index

Page numbers in bold refer to key locations for this information.

acoustic detection **192–4**
  acoustic tags 36, 193, 194, 198, 200
  acoustic tracking 36, 193
action bias 105
advanced warning system *see* early-warning system
aerial patrol *see* aerial surveillance
aerial surveillance 176, 183, **188–92**, 211
Aeromonas 146
Alopias vulpinus 22, *see also* common thresher shark
Amazon River 56
ampullae of Lorenzini 34, 35, 67, 117, 119
ampullary system *see* ampullae of Lorenzini
analytical system 101
anatomy **16–20**, 35
animal welfare 152, 194
anomalous weather 52, 57, 58
anoxia 10–12
antibiotic therapy 136, 137, 141, 145
antibodies, shark 10, 13
anxiety 70, 93, **97–100**, 148, 150
apex predator 1, 8, 168
app 110, 136, 168, 183, 199
Arctocephalus forsteri 37, 57, *see also* New Zealand fur seal
Arctocephalus pusillus pusillus 37, *see also* Cape fur seal
Atlantic Ocean 5, 37, 40, 56, 154
attack rate 157, 170, 202, 204, 218, 219
auditory cue 124, 125, 197
auditory deterrent **124–5**
Australia 5, 38, 47, 53, 56, **57**, 72, 83, 88, 154, 156, 158, 163, 168, 194, 215, 216, 217, 222
Australian Shark Attack File 47, 83

Bahama Islands *see* Bahamas
Bahamas 56, 57, 215
baiting *see* chumming
ban, activity 54, 108, 158, 167, **182–4**, 203, 205, 206, 214, **217–24**
basking shark 15, *see also Cetorhinus maximus*
beach mesh net 5, 165, 169, **173–7**, 181, 182, 190
beach net *see* beach mesh net
behaviour
  human 82, 99, 100, 102, 104, 159, 163, 211
  shark 7, 8, 12, 16, 19, 20, 30, 31, 34, 38, 39, 40, 42, 48, 52, 59, 79, 81, 111, 112, 121, 122, 124–5, 155, 169, 170, 196, 220, 222
bite force 20, **21**, 141
blacknose shark 49, 63, 121, 205, *see also Carcharhinus acronotus*
blacktip reef shark 12, 17, 112, *see also Carcharhinus melanopterus*
blacktip shark 5, 34, **41**, 49, 63, 145, 146, 180, *see also Carcharhinus limbatus*
blood loss *see* haemorrhage
blue shark 5, 122, *see also Prionace glauca*
boat traffic 59, 125, 193, 195
bony fish *see* teleost
brain
  human 11, 98, 100, 101, 102, 106
  shark 22, **30**
Brazil 5, 56, 57, **58–9**, 60, 125, 163, 173, 182, **202–5**, 218, 221
bronze whaler 198, 199, 200, *see also Carcharhinus brachyurus*
brown banded bamboo shark 11, *see also Chiloscyllium punctatum*
bubble curtain 197
buccal pumping 21, 22
bull shark 5, 7, 21, 28, 29, **40–1**, 52, 56, 59, 61, 62, 91, 114, 122, 124, 145, 157,